Membrane Bioreactors
Operation and Results of an MBR Wastewater Treatment Plant

STOWA in brief

The Foundation for Applied Water Research (in short, STOWA) is a research platform for Dutch water controllers. STOWA participants are all ground and surface water managers in rural and urban areas, managers of domestic wastewater treatment installations and dam inspectors.

The water controllers avail themselves of STOWA's facilities for the realisation of all kinds of applied technological, scientific, administrative legal and social scientific research activities that may be of communal importance. Research programmes are developed based on requirement reports generated by the institute's participants. Research suggestions proposed by third parties such as knowledge institutes and consultants, are more than welcome. After having received such suggestions STOWA then consults its participants in order to verify the need for such proposed research.

STOWA does not conduct any research itself, instead it commissions specialised bodies to do the required research. All the studies are supervised by supervisory boards composed of staff from the various participating organisations and, where necessary, experts are brought in.

The money required for research, development, information and other services is raised by the various participating parties. At the moment, this amounts to an annual budget of some 6,5 million euro.

For telephone contact number is: +31 (0)30-2321199.
The postal address is: STOWA, P.O. Box 8090, 3503 RB, Utrecht.
E-mail: stowa@stowa.nl.
Website: www.stowa.nl.

STOWA Report

Membrane Bioreactors
Operation and Results of an MBR Wastewater Treatment Plant

A.G.N. van Bentem, C.P. Petri, P.F.T. Schyns and H.F. van der Roest

IWA Publishing

Life

stowa

Published by IWA Publishing, Alliance House, 12 Caxton Street, London SW1H 0QS, UK
Telephone: +44 (0) 20 7654 5500; Fax: +44 (0) 20 7654 5555; Email: publications@iwap.co.uk
Web: www.iwapublishing.com

First published 2007

Translation by Katarzyna Kujawa-Roeleveld

British Library Cataloguing in Publication Data
A CIP catalogue record for this book is available from the British Library

Library of Congress Cataloging- in-Publication Data
A catalog record for this book is available from the Library of Congress

ISBN: 1843391732
ISBN13: 9781843391739

Contents

SUMMARY

The application of a membrane bioreactor (MBR) can represent an important leap forward in the treatment of municipal waste water. Compared to traditional treatment techniques, it requires a far smaller surface area and produces better effluent quality. Recently in the Netherlands, pilot-studies at various locations (WWTPs at Beverwijk, Maasbommel, Hilversum and Leeuwarden) gained experience with MBR technology. Research results have led to a decision to build an MBR demonstration plant at the Rhine and IJssel Water Board's WWTP in Varsseveld. This first full-scale MBR in the Netherlands began operation in December 2004. This report describes the results and experiences of the first 16 months of operation and the associated research program.

As a demonstration plant, the Varsseveld MBR must prove that MBR technology can be applied to the treatment of municipal waste water in the Netherlands. This means that MBR must meet the Netherlands' far-reaching effluent requirements target under stable conditions and for a reasonable cost. With this in mind, the following objectives were formulated for the study:

1. Establish technological results;
2. Optimise processes and demonstrate stable operation;
3. Evaluate the design and establish the need for redundancy;
4. Compare the costs of MBR with those of a conventional WWTP;
5. Demonstrate the technical feasibility of scaling up.

This report also employs this sequence in describing the results and experiences of the operation and the relevant research aspects. A summary of the most important findings is included below.

1. RESULTS

The biological results of the Varsseveld MBR are good. The relatively strict target value for nitrogen in the permeate (5 mg N_{total}/l) has been achieved comfortably and the target value for phosphate (0.15 mg P_{total}/l) can be achieved with a dosing of about 1.2 mol Fe/mol $P_{influent}$. Despite the lack of an anaerobic tank, biological phosphate removal plays an important role. Due to this, and despite fairly high concentrations of phosphate in the influent, relatively small volumes of chemicals are required.

The membrane plant has always functioned well, with the exception of the first couple of months, when the discharge of an industrial polymer interrupted operations. Since the membranes' capacity has been restored (by intensive cleaning of the membranes and the ceasing of the discharge), the plant has been able to process the influent under all conditions. Under relatively favourable conditions (high process temperature), the membrane guarantee tests were carried out successfully. In the first year, much attention was given to reducing the energy consumption of the membrane filtration plant in particular. Despite this, the Varsseveld MBR's energy consumption is still about 35% higher than that of a conventional activated sludge system with sand filtration. After a few alterations to the plant's controls, the forecast is that energy consumption will fall further, meaning that Varsseveld MBR's energy consumption will still be about 15% higher than that of a conventionally activated sludge system with sand filtration.

2. OPERATIONS

Reliable pre-treatment is essential for reliable membrane operation. Pre-treatment, especially of the micro-sieves, is the part of the Varsseveld MBR which has required the most attention. During the beginning and/or end of rain-weather conditions, the micro-sieves have often become blocked. Further optimisation of the micro-sieves should prevent these problems in the future. Most of the time, the membranes have functioned perfectly, both from a technical and a technological viewpoint. Permeability is still at the same level 16 months after the start up of the plant. Leakages in the membrane modules can be put down to scaling up the MBR technology to practical applications. These have been dealt with and remedied correctly by the supplier. Furthermore, macro-contamination of the membranes has been detected (hair and leaves at the top of the modules and chunks of sludge at the bottom). These problems are expected to have been sorted out after a number of technical alterations and further optimisations of the process settings.

3. DESIGN

Pre-treatment aims to protect the membranes, and the other process components, against contamination and wear. The fine screens and the sand/fat collector have always functioned without problem. However, the micro-sieves appear to be the part of the plant most sensitive to malfunction. Further to the experiences with and results of pre-treatment, the following design recommendations have been made:

- For the time being, a redundancy of 100% should be assumed for the micro-sieves and 50% for the fine screens, redundancy is not required for the sand/fat collection unit;
- A separate fat collector does not appear to be essential for the effective operation of the MBR. This requires further research.
- A separate sand removal system is recommended, though, in order to prevent damage to the process parts and the membranes.

The Varsseveld MBR's biological section consists of an aerated circulation system and a pre-denitrification tank. However, based on measurements of the oxygen input and the hydraulics of the activated sludge system, it can be concluded that a circulation system is not optimal for such an application. Because a relatively high amount of air must be introduced into a small volume, the space available for the aeration elements and propellers is limited, which leads to sub-optimal process conditions. Together with over-optimistic design principles, this appears to reduce oxygen input in the activated sludge system to about 35%

lower than predicted by the design. In the current situation, it appears that membrane aeration is responsible for 20-25% of oxygen input.

Design of the MBR's activated sludge system should assume removal in pre-treatment. Pre-treatment like that in Varsseveld, with a fine screen, a sand/fat collector and 1 mm micro-sieves in succession, allows the assumption of a 20-30% removal of COD, BOD and SS. For nitrogen and phosphate, this is 5-10%. The MBR appears to be more sensitive to the formation of a floating layer than to conventional activated sludge systems. This must be taken into account in the design.

On the basis of the guarantee tests, it can be concluded that the membrane filtration plant has met the design requirements. This concerns a design flux of 37.5 l/(m^2.h) over 3 days followed by a maximum flux of 50 l/(m^2.h) over 8 hours. However, the guarantee tests were carried out under conditions favourable to membranes, i.e. at temperatures above 20°C. On the basis of experience with peak loads in the winter, the design flux employed appears to be optimal for the process conditions at the Varsseveld MBR. An even higher design flux is not recommended for this type of membrane under these conditions.

The scaling up of the membrane tanks to large-scale applications appears to work well. The system's symmetry is proven and the membranes function extremely well. Further optimisation for future designs requires both increasing the membrane tank capacity and reducing the tank dimensions.

4. COSTS

The cost of setting-up the Varsseveld MBR was comparable roughly to that of expanding the capacity of a conventional sand filtration plant. The utilization costs (operational costs and capital charges) of the Varsseveld MBR are about 17% higher than those of the reference variant. Table A gives a summary of set-up costs and utilization costs for the Varsseveld MBR.

Table A. A summary of set-up costs and utilization costs for the Varsseveld MBR

Cost item	Unit	Varsseveld MBR	Reference variant	Difference
Set-up costs	€	11,200,000	11,200,000	-
Operating costs	€/year	1,640,000	1,400,000	240,000

This cost comparison suggests that the cost differential between an MBR and a conventional WWTP with sand filtration is relatively limited. Future developments (lower energy consumption, lower membrane prices) may lead to a further reduction in the MBR utilization costs.

5. TECHNICAL FEASIBILITY

The first 16 months' operational results and experience with the Varsseveld MBR suggest that MBR technology is suitable for the treatment of municipal waste water in the Netherlands. The plant's results have in general been very positive. Target values for permeate quality (5 mg N_{total}/l and 0.15 mg P_{total}/l) have been achieved and membrane function is excellent (apart from a few technical problems). The costs of the MBR are already comparable to those of a conventional plant with sand filtration, and the cost differential is now minimal with respect to energy consumption.

The two most important critical factors that could thwart the MBR's success are pre-treatment and influent composition:

- A stable and reliable generation of microsieves requires development;
- Greater understanding is required concerning membrane sensitivity to components present in the waste water.

Finally, for the further development of MBR technology, fundamental scientific research into, amongst other things, the relationship between sludge quality and filterability is essential.

1

Introduction

1.1 INTRODUCTION

The implementation of the membrane bioreactor (MBR) can represent an important step forward in the treatment of municipal wastewater. Compared to the traditional treatment techniques it has a smaller foot print and produces better effluent quality. Recently in the Netherlands experience with MBR technology has been gained during pilot studies carried out at various locations. In the year 2000 there was a large pilot research study started on the WWTP Beverwijk into the feasibility of this new technology [1,2]. In addition to these fast technical and technological developments a strong price decrease of the membranes on the world market contributed to the introduction of the MBR-technology. In the Netherlands the cooperation between STOWA, various water boards, membrane manufacturers, knowledge institutes and consulting companies was established. There was also a strategy worked out for MBR development in the Netherlands where a prominent place has been reserved for a demonstration installation (Figure 1.1). This demonstration installation is an important step towards the realisation of full-scale MBR plants. The WWTP Varsseveld fits perfectly with respect to its size and retrofit plans. Water Board Rijn en IJssel supported by financial contributions from the European Commission (LIFE) and the innovation funds from the STOWA decided to build the first full scale MBR installation in the Netherlands to treat domestic sewage.

For the successful introduction of a large scale MBR system the technical aspects of the scaling-up are very important. This is why much attention was paid to the demonstration installation of the design of the standard membrane tanks (units), the process control and the optimisation and automation of membrane cleaning. The membrane units after their functioning has been proved can be installed at larger scale MBR installation. That is why, for the start-up and the first year of operation an extended research program was included. This research program was started following agreement between all involved partners: the Water Board Rijn en IJssel (WRIJ), STOWA, DHV and members of the advisory board. The goals of the research are described in §1.2. The research program consists of eight sub-studies. The set-up and the organisation of the research program is explained in §1.3.

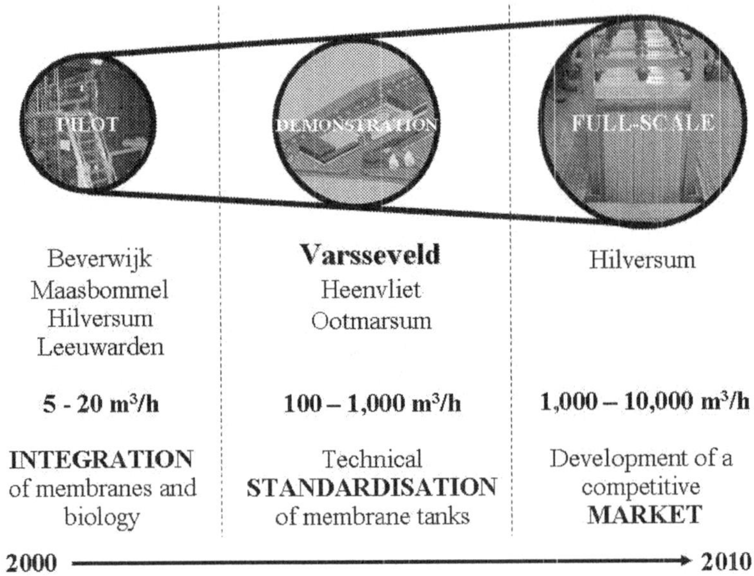

Figure 1.1: The development pathway of the MBR

The results of the first year of the operation of the MBR Varsseveld and the research programs are described in two STOWA reports. This comprises this report and a sub-studies report [3]. Guidance for this main report is given in §1.4.

1.2 RESEARCH OBJECTIVES

The main objective of the MBR Varsseveld, as a demonstration plant, was to prove that the MBR technology is applicable for the treatment of domestic wastewater in the Netherlands. This means that the MBR has to comply with effluent requirements in the Netherlands at stable conditions and at reasonable costs. Based on this starting point the following research objectives were formulated:

(1) Demonstration of the technical feasibility of the scaling-up
(2) Demonstration of a stable operation
(3) Collection of the technological results, divided into pre-treatment, biology and membranes
(4) Research into the necessity of redundancy
(5) Optimisation of the operation
(6) Evaluation of the design
(7) Comparison of the costs of an MBR and a conventional WWTP.

The most important aspects named in the objectives were design, operation, technological results and costs. An overview of these aspects and the relations between them are given in Figure 1.3. In this figure the numbers represent the place of the objective within this scheme.

Figure 1.2: The WWTP Varsseveld with the new MBR (right) and an old installation taken out of operation (left)

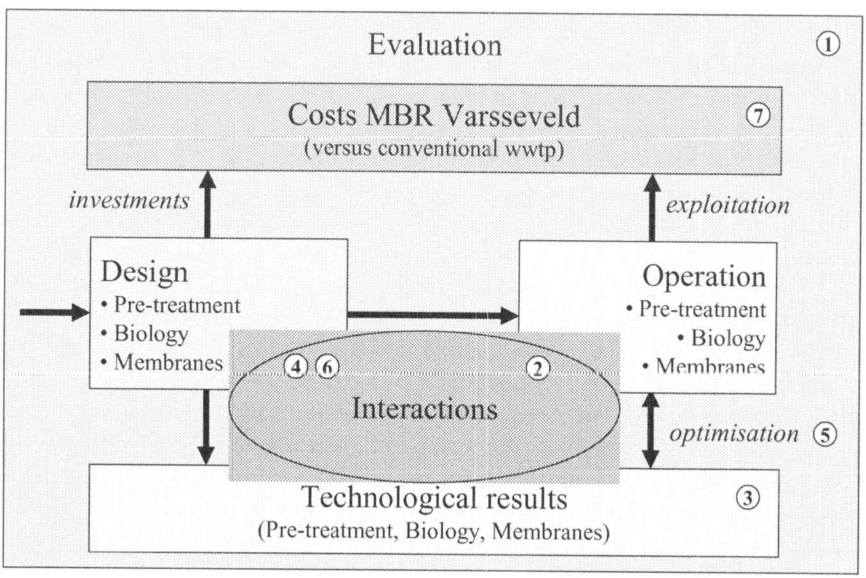

Figure 1.3: Relationship between different objectives of the research Varsseveld

The design of the MBR Varsseveld is based on the technical standards and knowledge of 2001. The experiences of the research in Beverwijk [1,2] and the available information on existing MBR installations abroad were used. The operation of the installation was optimised in the first year, with a main goal to improve the working of the installation and to lower the consumption of chemicals and energy. The design and the way of operation lead together to the achieved technological results. The costs of the installation are the result of both, the design (investment) and the operation (exploitation).

The relationship between design, operation and results enables the identification of potential improvements in design and operation. The evaluation as a whole needs to lead finally to an answer to the question of whether MBR is a serious alternative for the treatment of domestic wastewater in the Netherlands.

1.3 RESEARCH SET-UP AND ORGANISATION

Based on these objectives the research program was formulated. This program was divided into eight parts. These eight sub-studies are given below together with a description of the objectives and motivation:

1. *Operation and effluent quality.* An MBR comprises a number of sub-processes where no experience has been gained until now in the Netherlands. This includes, for instance, pre-treatment and membranes. From experience with operation of a similar new installation lessons for the future should be learned. The main attention in this sub-study was paid, amongst others, to the effluent quality and the scum formation.
2. *Pre-treatment*: The required degree of wastewater pre-treatment is higher than needed for a conventional WWTP and is of great importance for good membrane performance. The functioning of the pre-treatment is therefore one of the most critical aspects. There is however still little experience with this type of pre-treatment in full-scale. The most important objective of this sub-study was the determination of the removal percentages in the subsequent pre-treatment steps.
3. *Removal specific pollutants*: The New Water Framework Directive indicates compounds normally not monitored during a domestic wastewater treatment. This comprises for instance hormones, pharmaceutical residues and pesticides. As continuation of the previous STOWA studies [4] where the efficiency of different treatment techniques to remove these specific micropollutants was evaluated, a monitoring program was also performed in the MBR Varsseveld.
4. *OC and hydraulics*: The compact character of the MBR leads consequently to a larger demand for aeration capacity than the conventional WWTP while the volume to aerate is significantly smaller. This has an effect on the oxygen transport and the hydraulics of the aeration tank. The goal of this study was to determine the oxygen transport and the flows in the aeration tank and the membrane tanks for the different specific operational situations.
5. *Sludge quality versus filterability*: It is known that the performance of membrane filtration depends strongly on the sludge quality. There is however little known

about the relation between sludge properties and filterability of the sludge. The goal of this sub-research was to get a better understanding of these complex interactions and based on them to optimise the operation of biological treatment processes.

6. *SIMBA modelling*: The dynamics of an MBR installation differ strongly from that of a conventional WWTP. The hydraulic retention time in the aeration tank is significantly shorter and the buffering property of the secondary clarifiers is absent. The goal of this sub-study was to optimise the process control of the MBR Varsseveld and the evaluation of the design using dynamic simulation.
7. *Membranes:* One of the most important goals of the demonstration installation is the scaling-up of the membrane tanks. The membrane tanks, as implemented in Varsseveld, are considered as standard units for larger installations. Attention was paid in this research to, amongst others, the hydraulic functioning of these tanks and further optimisation of the process operation of the membranes.
8. *Simulation unit*: The world-wide experiences gained with the MBR technology proved that a simulation unit can significantly simplify and accelerate a technological start-up and the operation of the MBR installation. Moreover a simulation unit on Varsseveld wastewater forms important support for the practical scale research. Changes in process can be first tested using a simulation unit before being applied in the large scale installation. This allows not only lowering of costs but also minimising the operational risks associated with the operation of the full scale installation. In addition a simulation unit offers the possibility to train the plant operators of the Water Board.

The investment and operational costs of the MBR are higher than those of a conventional WWTP. To cover the surplus costs, STOWA supported by all Dutch Water Boards, called an Innovation Foundation into being. This foundation was established to give a positive impulse to technological developments such as the MBR in Varsseveld.

The broader significance of the project was also recognised by the European Commission and two ministries: Housing, Spatial Planning and the Environment (VROM) and Economic Affairs. In this way a financial contribution of the European Commission in the framework of the financial instrument LIFE was admitted to the project. The minister of VROM acknowledged this project as important for environmental protection and declared it as 'green'. The minister of Economic Affairs granted a subsidy within the framework of the Dutch Subsidy Regulation for energy supplies in non-profit and exceptional sectors.

1.4 STRUCTURE

The results of the various sub-studies were reported in two STOWA reports. An overview of the different studies and the structure of the report is given in Figure 1.4. This book comprises the main report where the most important results were reported. The structure of this main report is based on the objectives of the research project, as schematically shown in Figure 1.3. In Chapter 2 an overview is given of the design of the WWTP Varsseveld. For each process unit the design, the set-up and the operation are described. The technological results of the installation are presented in Chapter 3. In Chapter 4 the operation of various process units is described; the main attention is paid to optimisation in the first year of operation. The relation between the design, the operation and the results is described for a number of important aspects described in Chapter 5. The costs of the MBR Varsseveld were calculated and compared to the costs of a conventional extension of the WWTP Varsseveld in Chapter 6. Various aspects are evaluated in Chapter 7. The glossary is given in Appendix 1.

<table>
<tr><td rowspan="2">Report 1 – Main report</td><td>Presentation and evaluation of all research activities</td></tr>
<tr><td>Side study Operation en effluent quality</td></tr>
<tr><td rowspan="7">Report 2 – Side studies</td><td>Side study Pre-treatment</td></tr>
<tr><td>Side study Removal of micro-pollutants</td></tr>
<tr><td>Side study OC and Hydraulics</td></tr>
<tr><td>Side study Sludge quality versus filterability</td></tr>
<tr><td>Side study SIMBA-modelling</td></tr>
<tr><td>Side study Membranes</td></tr>
<tr><td>Side study Simulation Unit</td></tr>
</table>

Figure 1.4: Structure of the MBR reports

2

Design MBR Varsseveld

2.1 INTRODUCTION

The wastewater treatment plant Varsseveld was a conventional high loaded activated sludge system not able to comply with the current effluent requirements. The Water Board Rijn en IJssel decided in 1999 to extend the WWTP Varsseveld. Except for enlarging the biological capacity and sharpening the effluent requirements regarding nitrogen and phosphate removal, the prevention of accidental sludge wash out, the reduction of odours and noise nuisance were important objectives. The effluent is discharged to a vulnerable small inland waterway, Boven Slinge, an ecological connection zone.

Figure 2.1: Inland waterway Boven Slinge

The design of the MBR Varsseveld is based on the technical standard and knowledge of 2001. The experiences from the research in Beverwijk and the available information about existing MBR installations abroad were mainly used. The most important design data of the MBR Varsseveld are reported in this chapter. The basic assumptions used for the design are summarised in §2.2. Subsequently, the designs of the pre-treatment, the biology and the membrane tanks are described in §2.3 to §2.5. Attention was hereby paid to the dimensioning, the functional operation and the process control.

2.2 DESIGN ASSUMPTIONS

2.2.1 Introduction

The wastewater from residential areas (Varsseveld, Mariënvelde, Heelweg, Halle and Westendorp) is transported to WWT Varsseveld via three pressure mains. The MBR is designed for a pollution load of 23,150 p.e. (à 54 g BOD/d) and a maximal hydraulic capacity of 755 m^3/h. An important contribution of the loading (approximately 30% of the loading) originate from a local cheese factory. In this paragraph the most important assumptions are reported. §2.2.2 gives an overview of the influent characteristics and the effluent requirements, while in §2.2.3 the outcomes of the risk analysis are described. This was performed in the design phase in order to predict the risks of the design and, when possible and demanded, to minimise them.

2.2.2 Process parameters

The design of the MBR Varsseveld was based on the starting points reported in Tabel 2.1.

Table 2.1: Design parameters WWTP Varsseveld

Parameter		Unit	Value
Designed loading	Population equivalent	p.e. (á 54 g BOD)	23,150
		p.e. (á 136 g TOD)	31,470
Hydraulic loading	Average daily flow	m^3/d	5,000
	Dry weather flow	m^3/h	250-300
		m^3/d	3,500
	Rain weather flow	m^3/h	755
Pollutants load	Chemical oxygen demand (COD)	kg/d	3,000
	Biological oxygen demand (BOD_5)	kg/d	1,250
	Kjeldahl nitrogen (N_{Kj})	kg N/d	280
	Phosphate	kgP/d	50
	Suspended solids (SS)	kg/d	1,375

The treated wastewater is discharged to the inland waterway Boven Slinge. The effluent requirements are summarised in Table 2.2. There is a distinction made between the legal requirements and target values defined by the Water Board. These target values are valid in the periods where the receiving body of surface water is the most vulnerable and sensitive to eutrofication, which is in the months from April until the end of October.

Table 2.2 Effluent quality requirements

Parameter		Unit	Requirement	Target value[4)]
Chemical oxygen demand	(COD)	mg/l	≤ 100[1)]	-
Biochemical oxygen demand	(BOD)	mg/l	≤ 10[1)]	-
Nitrogen	(N_{total})	mgN/l	≤ 10[2)]	≤ 5
Phosphate	(P_{total})	mgP/l	≤ 1[3)]	≤ 0.15
Suspended solids	(SS)	mg/l	≤ 10[1)]	≤ 5

[1)] daily average concentration

[2)] yearly average concentration from volume proportional samples

[3)] average concentration of 10 subsequent volume proportional samples

[4)] the target value is applicable in the period April till the end of October; the evaluation is based on the average concentration of volume proportional samples over the period April till the end of October.

The minimum process temperature at the WWTP Varsseveld was set to 7.5°C. Design temperature is important for both, biology and the membrane installation. The minimum temperature determines, amongst others, the sludge loading; for membrane installations the design flux is strongly dependent on the minimum temperature.

An important design criterion is the available space. Because it was the first full scale installation in the Netherlands it was decided to keep the existing conventional installation as back-up. This installation was taken out of operation but not removed, as shown in Figure 1.2. This old installation can be put in operation again, when necessary. After the MBR proved to function satisfactorily, the old installation will be dismantled. As shown in Figure 2.1, the available space for MBR was restricted. The installation (pre-treatment, biology, membrane tanks, pumps- and blower room and air treatment) was built on a surface of 85 x 21 meters. The sludge buffer, sludge thickener and ferric dosing facilities were built next to this block. The existing operational building was extended and modernised.

2.2.3 Risk analysis

The MBR Varsseveld is a demonstration installation serving as a model for the further development of the MBR technology in the Netherlands. Therefore it was important to consider the critical elements of the design and where possible to overcome the problems. For this reason a risk analysis was performed by the Water Board Rijn and IJssel and DHV to form a base for the design. The process engineers and operators of the WWTP Varsseveld were both involved in this process.

The risk analysis performed by the design team focussed on identification and quantification of the risks, and where possible, the measures were indicated to minimise the risks. For each risk it was estimated how big was the probability that it would occur and what kind of consequences there could be. Only in a case when the probability of occurrence and the consequences were significant, were the additional measures defined. The most important design precautions derived from the risk analysis are given below:

- The fine screens and the micro-sieves were made with 100% redundancy;
- For the design of aeration a sufficient safety was taken into account. The α-factor of 0.5 was assumed;

- The by-pass to the old WWTP can be used under certain circumstances;
- The capacity of the membrane filtration can be relatively easily increased. The capacity of each membrane cassette can be increased with 20% by placing new elements. The aeration capacity has then enough flexibility to cope with this increase. In addition, in each membrane compartment, a fifth cassette can be placed.

Results of the MBR Varsseveld gave insight into the necessity for these additional provisions. The attained understanding allows the optimisation of the design of future MBRs.

2.3 PRE-TREATMENT

2.3.1 Introduction

Because membranes are sensitive to fouling and damage, much attention has been paid in the design to the pre-treatment of the raw wastewater. It is important to remove leaves, plastic, sand, fat, hair and similar debris from the wastewater. Therefore the wastewater flows first through a fine screen (§2.3.2) with a bar opening of 6 mm, followed by an aerated grit- and grease removal (§2.3.3) and finally micro-sieves with a perforation of 0.8 mm (§2.3.4). There is no by-pass of the fine screens and micro-sieves. Even with such a degree of pre-treatment, solid particles can enter the activated sludge system and impede the operation of the membranes or damage the membranes. Falling leaves or dry sludge particles can cause these problems. To keep the activated sludge clean the continuous internal recycling from the activated sludge tank passes the micro-sieves. Moreover, during a membrane cleaning cycle, when the membrane tank is emptied, the whole content of this tank is recycled through the micro-sieves (Figure 2.2). The discharge of the scum layer is further described in §2.4.5.

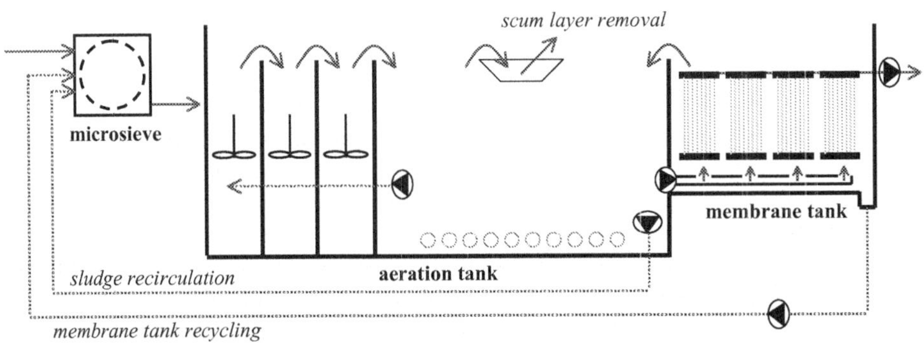

Figure 2.2: Schematic representation of the sludge recycling and scum layer withdrawal

2.3.2 Fine screens

The wastewater from the different pressure mains is received in a receiving chamber, from where it is sent to the fine screens. From the fine screens with a clear spacing between the bars of 6 mm, the debris are removed from the wastewater (Figure 2.3 left). Because of the great importance of a good working pre-treatment the fine screens were made very robust. The run off water from the treatment site was brought to the fine screens. The screenings were transported with a screw pump to the dewatering/press device where they were washed and dewatered. After, the screenings were stored in a container (Figure 2.3, right)

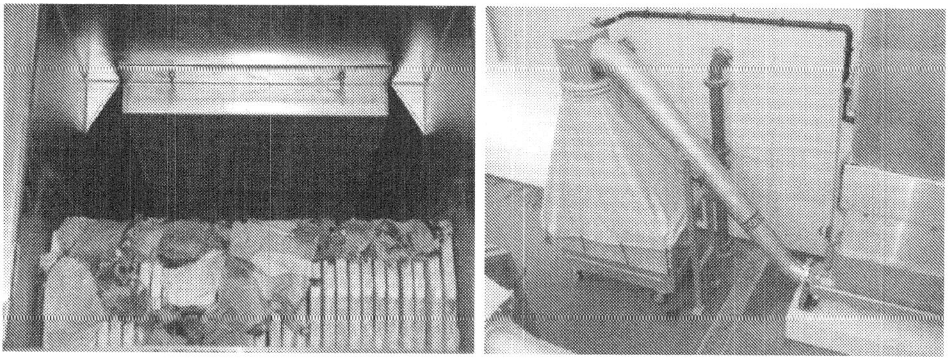

Figure 2.3: Screening of screenings material

Each of the two fine screens with a belonging transport screw pump forms an independent screen goods removal system. Based on the number of operational hours, once a week the screen goods removal system is switched from one to the other. The step screen switches on based on the water level before the screen. After exceeding the (adjustable) switch level, the screen switches to another step. This is repeated until the level is again below the switch stand. By exceeding the high water level, the screen operates continuously during a set time.

2.3.3 Grit/grease removal

In order to minimise the sand deposition in the lower volumes of the installation and the damage and contamination (fouling) of the membranes caused by sand and fat, the wastewater is led through a combined grit and grease removal system. In this sand/grease removal system the flow velocity of wastewater is selected in a way that the most heavy particles (namely sand) will settle. The settled sand accumulates in a slit in the bottom of the grit removal tank. Due to implementation of coarse bubble aeration the screw shape flow pattern is created causing flotation of fat. The settled and the floating material are removed using a conveyor with scrapers and flights to remove the scum layer (Figure 2.4 left). The sand is pumped from a slit using a sand pump and transported to the grit washers (Figure 2.4, right). In the grit washer, the settled material is washed with a run off water using an intensive mixing. The washed product consists of a large fraction of sand. The scum layer is discharged to the grease sump together with the floating layer from the aeration tank. The content of the grease sump is regularly collected by a truck.

Figure 2.4: Photos of the grit chambers (left) and the grit washer (right)

2.3.4 Micro-sieves

The micro-sieves with round openings of 0.8 mm allow the removal of the fine particles (e.g. hair) from the wastewater and thus protect membranes against fouling and damage. Because of the large importance of a trouble free working pre-treatment the microsieve streets are made robust. There are two microsieve streets, each with its own sieve goods conveyer and transport pump. Because the MBR Varsseveld is the demonstration plant, both micro-sieve trains are equipped with different sieve types:

- Type 1: rotating drum sieve with a sprinkle system from the Auxill company (Figure 2.5, left)
- Type 2: stationary sieve with brush cleaning of the Contec company (Figure 2.5, right).

Except for the wastewater from the grit chamber, also the sludge is recycled over the micro-sieves. This sludge is recycled sludge from the aerated circuit (only at dry weather conditions). In this way the activated sludge in the system is kept as clean as possible. Besides, the sludge pumped from the membrane tanks after the chemical cleaning of the membranes is also sieved. The sieve goods from the Auxill drum sieves are removed with a sprinkle from the inner surface of the drum. The sieve goods from the stationary Contec sieves are removed by rotating brushes. The sieve goods are pumped with sieve goods pumps to the sludge buffer tank. It is mixed with the thickened excess sludge and transported with a truck.

The two micro-sieve streets, including sieve goods conveyor and sieve goods pump serve as reserve in relation to each other. At the set time the micro sieve street is switched to the other one. There is always at least one micro sieve street in operation. During change over from dry weather to storm weather flow both streets are in operation.

Type 1 – Rotating drum sieve with sprinkle installation

Type 2 – Stationary sieve with brush cleaning

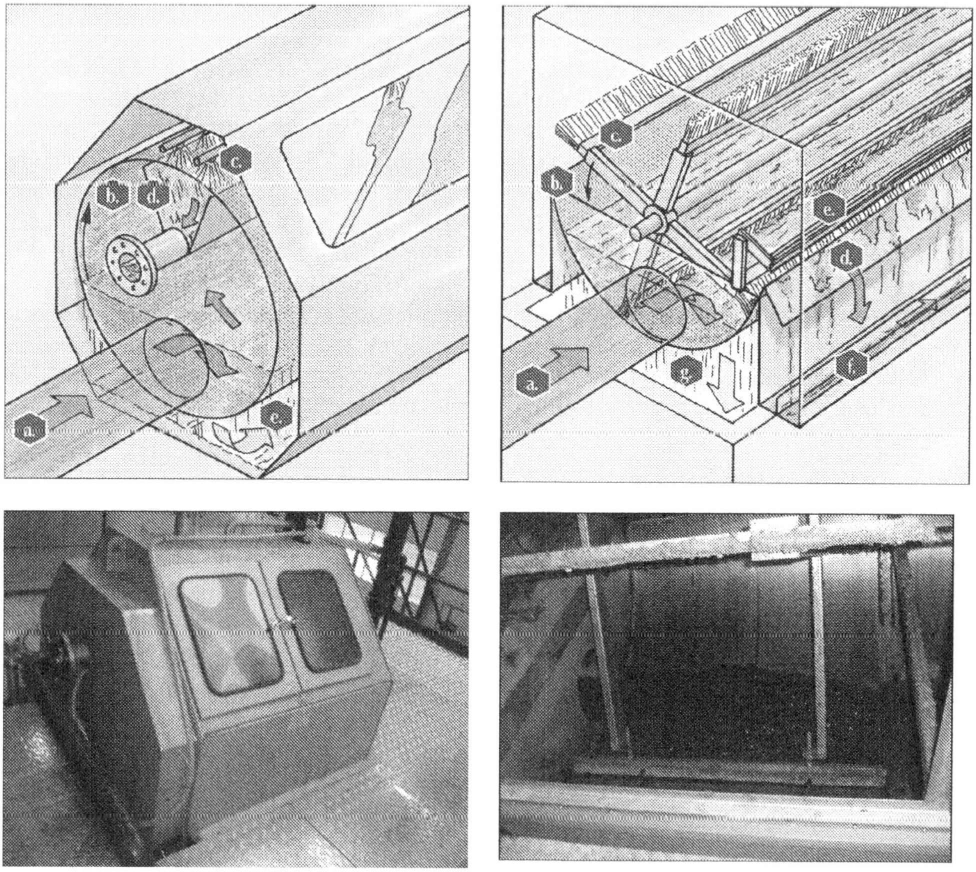

Figure 2.5: Pictures and schemes of the micro sieves

2.4 BIOLOGY

2.4.1 Introduction

The biological processes taking place in an MBR are in principle not different from those in a conventional activated sludge system. Since a higher sludge concentration can be applied in an MBR and the secondary clarifiers are not needed, an MBR is more compact. The design of the biological treatment and the control of the biological processes require another approach because of the shorter hydraulic retention time and the high oxygen concentrations

in the membrane tanks. Therefore the SIMBA model assisted the design and process control. Based on the SIMBA model, amongst others, it is concluded that to reach an effluent quality of 5 mg N_{total}/l in the MBR Varsseveld, an aeration circuit (with a high recycle rate) with a separate pre-denitrification had a preference.
In addition, the sludge characteristics played an important role in the design. Although the sludge characteristics in an MBR does not have direct impact on the sludge separation (there is no settling step), the performance of the biology influences the operation of the membranes. A good sludge structure, a sludge floc with little filaments, little free swimming cells and little extra-cellular polymeric substances (EPS) lead to a decrease of the membrane fouling. This has consequences for the design of an MBR. EPS is mainly formed under 'stress' conditions. These can be created under, for instance, low oxygen concentration and too high energy input. This means that the oxygen input needs to have enough reserve and flexibility and that for mixing and pumping of the sludge only equipment with low speed engines are implemented. The design of the aeration tank and the aeration are described in §2.4.2 and §2.4.3 respectively. The most important process control strategies are described in §2.4.4. Finally, the discharge of the floating layer is described in §2.4.5.

2.4.2 Aeration tank

The aeration tank is configured as a circuit system with a pre-denitrification volume. The wastewater influent to the first compartment of the pre-denitrification tank, is made as a plug-flow system. From the pre-denitrification tank the water/sludge mixture flows to the aeration tank. The sludge is recycled from the aeration circuit to the pre-denitrification tank. The sludge/water mixture is pumped from the circuit using four membrane influent pumps. The activated sludge is separated from the treated wastewater via membrane filtration. The 'thickened' activated sludge is recycled via an overflow drain back to the aeration circuit. The configuration of the aeration tank is given schematically in Figure 2.6.

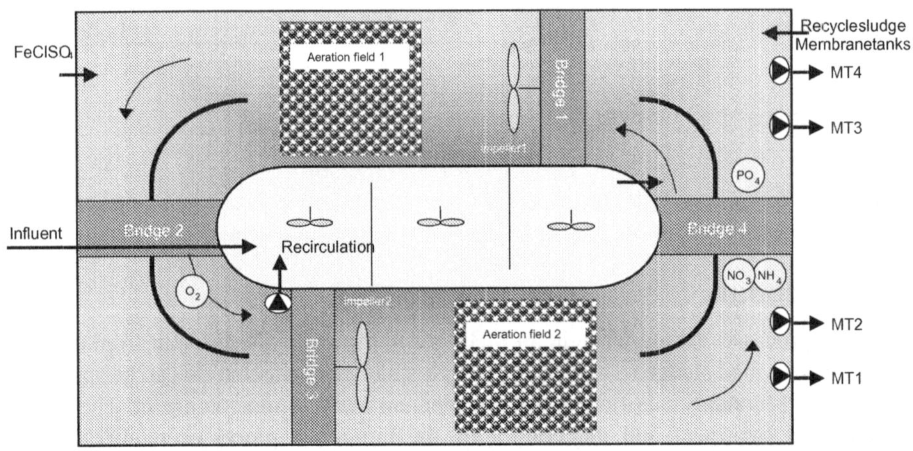

Figure 2.6: Schematic representation of the aeration tank

In the design special attention was paid to the hydraulics of the aeration circuit. Because the volume of the tank is relatively small, the distance between the impellers and the aeration fields and between the aeration fields and the curves of the circuit are also smaller than normal. This has consequences for the flow profile, the stream velocity and the oxygen input.

For chemical phosphate removal, an iron salt is dosed to the pre-denitrification tank. The choice for chemical phosphate removal and not biological with use of an anaerobic tank, was determined by way of the excess sludge handling process. The excess sludge of the WWTP Varsseveld is gravitationally thickened and transported to the other WWTP (Lichtenvoorde). With high efficient biological phosphate removal, a fraction of the uptaken phosphate could be released in the sludge line and it would contribute to the additional phosphate loading on the WWTP Varsseveld and/or on the other WWTP where the sludge is further handled. The most important dimensions of the aeration tank are given in Table 2.3.

Table 2.3: Dimensions of the aeration tank

Parameter	Value	Unit
Sludge loading rate[1)]	100	gCOD/kgMLSS/d
	42	gBOD/kgMLSS/d
	9.3	gN_{Kj}/kgMLSS/d
Sludge content aeration circuit		
total (biological/chemical)	10.0 (8.7/1.3)	$kgMLSS/m^3$
Iron salt dosing		
Summer/winter[2)]	2.0/1.0	$molFe/molP_{influent}$
Sludge production[3)]		
Total (biological/chemical)	1,530 (1,330/200)	kgMLSS/d
Sludge retention time[3)]	24	D
Aeration tank volume		
Total	3,000	m^3
Pre-denitrification	680	m^3
Circuit	2,320	m^3
Aeration tank depth	5.0	m
Recycle sludge pump	1,000-2,000	m^3/h

MLSS, mixed liquor suspended solids

[1)] calculated based on the volume of the aeration tank (excluding the membrane tanks)

[2)] The dosing is dependant of permeate quality regarding phosphate (Table 2.2)

[3)] Value obliging in the winter period

2.4.3 Aeration

The required specific aeration capacity, expressed in amount of air per hour and per cubic metre of the reactor, is significantly higher in an MBR than in a conventional activated sludge system. The reason for this, is a higher sludge concentration in an MBR. The sludge concentration in the MBR Varsseveld is 10 kg $MLSS/m^3$ while in a conventional system it is in the range of 4 $kgMLSS/m^3$, an increase of a factor of 2.5. The volume of the aeration tank is therefore approximately 2.5 times smaller. In addition the α-factor decreased at a higher sludge content, causing necessity of a higher air demand. Also as a result of the shorter hydraulic retention time, the peak factor is increased in the design stage. In total these three effects lead to an increase of the specific air inflow with a factor of 4. To express this in another way, approximately 60% more air needs to be provided to 40% of the volume.

In order to guarantee the above, an efficient way of aeration is necessary. On one side the system needs to be characterised by a good oxygen transfer, on the other side the space requirement for the aeration has to be limited. For these reasons, plate aeration has been chosen. The design parameters of the aeration are given in Table 2.4.

Table 2.4: Design parameters of the aeration

Parameter	Value	Unit
Maximal oxygen demand	273	kgO_2/h
Aeration height	4.9	M
Specific oxygen input	27	$gO_2/Nm^3/m$
Required capacity	2,080	Nm^3/h
Installed capacity	2(+1)*1,200	Nm^3/h

The installed aeration capacity is 2,400 Nm^3/h; the over capacity is 15% in relation to the required aeration capacity (see §2.2.3).

2.4.4 Process control

The aeration control of the MBR Varsseveld is based on nitrogen and oxygen levels. The oxygen set-point is determined by the ammonium and the sum of ammonium nitrate ratio. The on-line analysers placed in the aeration circuit are used. The cascade control is schematically given in Figure 2.7. When the required aeration demand is between 0-50%, only the first aeration field (Figure 2.6) is aerated. When the required aeration demand is above 50%, the air is redistributed between both aeration fields. In case of the malfunction of one of the analysers, the control switches to a direct oxygen measurement.

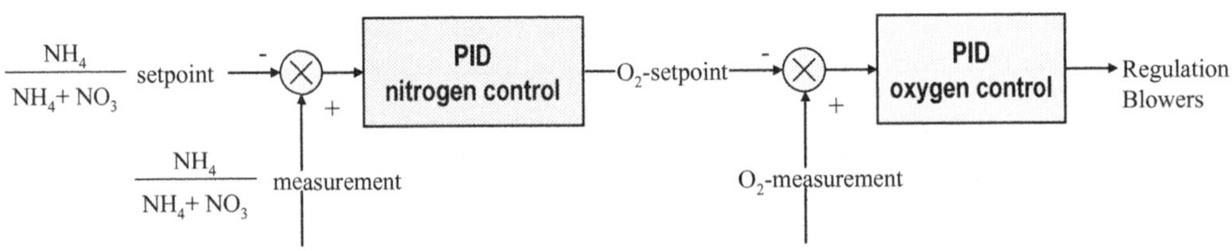

Figure 2.7: Schematic representation of the aeration control in the MR Varsseveld

The control of the internal sludge recycle to the pre-denitrification tank takes place based on the on-line nitrate measurement in the aeration circuit. The control strategy is shown schematically in Figure 2.8. Based on a low and a high nitrate set-point the recycle flow is controlled between maximal and minimal recycle flow. Hereby the recycle flow is lowered when the nitrate concentration in the circuit decreases. In this way the amount of nitrate brought to the pre-denitrification tank is kept on the same level.

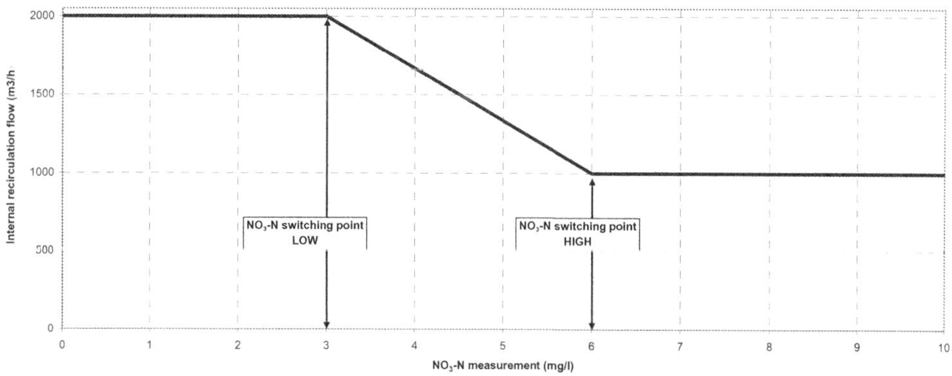

Figure 2.8: Schematic representation of the control strategy for internal recycle rate

The dosing of the iron salt can take place on the MBR Varsseveld in the influent to the pre-denitrification tank of the aeration circuit. The dosing rate can be controlled in three ways:

- Manually, with a constant dosing rate
- Based on the constant ratio in relation to the influent flow
- Based on the phosphate content in the aeration circuit.

The excess sludge is pumped from the aeration circuit to the sludge thickener. The pump is controlled based on the mixed liquor suspended solids (MLSS) concentration in the circuit. The pump is switched on in an on/standby function when exceeding a high set point (e.g. 10.5 kgMLSS/m^3) and is switched off after exceeding a low set point (e.g. 10.0 kg MLSS/m^3).

2.4.5 Scum layer discharge

In an MBR a scum layer will occur more probably than in a conventional activated sludge system because the floating material is not removed via the scum discharge system in the secondary clarifiers. Therefore to remove the floating material from an MBR an extra facility needs to be installed. The floating layer can consist of the bulking sludge and/or greasy material. This floating material has to be prevented from reaching the membrane. It is therefore important to remove the floating layer from the activated sludge tank and the influent from the activated sludge tank to the membrane reactors has to take place under the water level. The scum (or foam-) layer possibly formed in the membrane tanks need to be immediately brought back to the aeration tank. Because of these reasons the recycle from the membrane tank to the aeration tank takes place via an overflow on the surface level.

The above mentioned design strategy, applied in Varsseveld, is schematically given in Figure 2.2. Hereby the sludge recycle through the micro-screens is shown. The scum layer discharge is shown in Figure 2.9; it consists of a half static drum from where the scum layer is removed with brushes to the sump. The discharge of the scum layer is adjustable in height.

Figure 2.9: The scum layer discharge from the aeration circuit

2.5 MEMBRANE FILTRATION

2.5.1 Introduction

The membranes on the WWTP Varsseveld were provided by Zenon manufacturer; type ZW500d was installed (Figure 2.10). These are vertically installed capillary membranes with a pore size of 0.035 μm. The choice for Zenon membranes was based on the results of European tendering of the membrane filtration units, regarding membrane filtration units and the good experiences gained in, amongst others, WWTP Beverwijk. More detailed information regarding the design of membrane installation can be found in the sub-study 'Membranes' [3]. In this chapter the most important design data are described in §2.5.2. The set-up and configuration of the membrane tanks is discussed in §2.5.3. The control of the membrane installation is completely automised. The most important process control strategies for membrane filtration are described in §2.5.4.

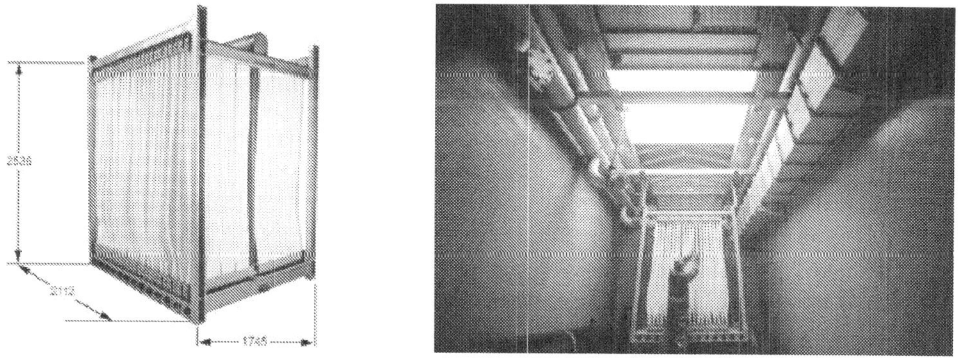

Figure 2.10. Scheme of a membrane cassette (left) and a picture of a cassette in a membrane tank.

2.5.2 Design parameters

The most important dimensions of the membrane installation on the WWTP Varsseveld are given in Table 2.5.

Table 2.5: Dimensions of the membrane installation on the WWTP Varsseveld

Process unit	Parameter	Value	Unit
Membranes	Type	Zenon ZW500d	-
	Design flux (net), rain weather flow	37.5	$l/(m^2.h)$
	Total surface	20,160	m^2
	N° cassettes	16	-
	N° elements per cassette	40	-
Membrane tanks	Number tanks	4	-
	Volume of each tank	170	m^3
	N° cassettes per tank	4	-
Membrane influent pumps	N° pumps	4	-
	Capacity of each pump	400 - 800	m^3/h
Permeate pumps	N° pumps	4	-
	Capacity of each pump	50 – 300	m^3/h
Membrane aeration	Total capacity	9,000	Nm^3/h
	Depth aeration tubes	2.8	m

An important design parameter of the membrane is the flux, which is the amount of water that can be filtered through a square metre of membrane in a time unit. The design of the MBR Varsseveld is based on the experiences from the Beverwijk research during a period of more than 4 years [1] and (industrial) full scale experiences elsewhere. The design flux of the MBR Varsseveld amounts to 37.5 $l/(m^2.h)$ net and is reached when all four membrane tanks are in operation during a storm weather flow. In the design it was assumed that this situation can last for maximally 3 days after each other. In case one the membrane tanks cannot operate during a storm weather flow (e.g. because of a technical malfunction), for a period of maximally 8 hour a net flux of 50 $l/(m^2.h)$ can be applied.

The design flux of 37.5 $l/(m^2.h)$ net is relatively high in relation to other existing installations abroad. For instance in Germany there is a maximum flux of 25 to 28 $l/(m^2.h)$ recommended, but in practice usually a lower design flux is used. The high design flux of the MBR Varsseveld results from the positive experiences in Beverwijk, where these high fluxes could be attained together with a good sludge quality. Here it has to be added that the storm-dry weather flow ratio in Varsseveld is relatively high in relation to the installations abroad. This has consequences that the maximum flux on the WWTP Varsseveld is perhaps high, but the average flux during dry weather conditions is on a similar level compared to installations in other countries.

2.5.3 Process operation

The membranes are placed in four separated compartments. For good operation of the membranes as equal a possible (symmetrical) loading is recommended. That is why special attention was paid to the design of the in- and out-flow of these tanks. The compartmentalisation makes it possible to take a membrane out of operation in case of low influent, when it is not necessary to use a full capacity. This allows energy to be saved, because membranes do not need to be continuously aerated. Also, the membranes then have the time to rest (relax). The set-up of the membrane tanks is schematically given in Figure 2.11.

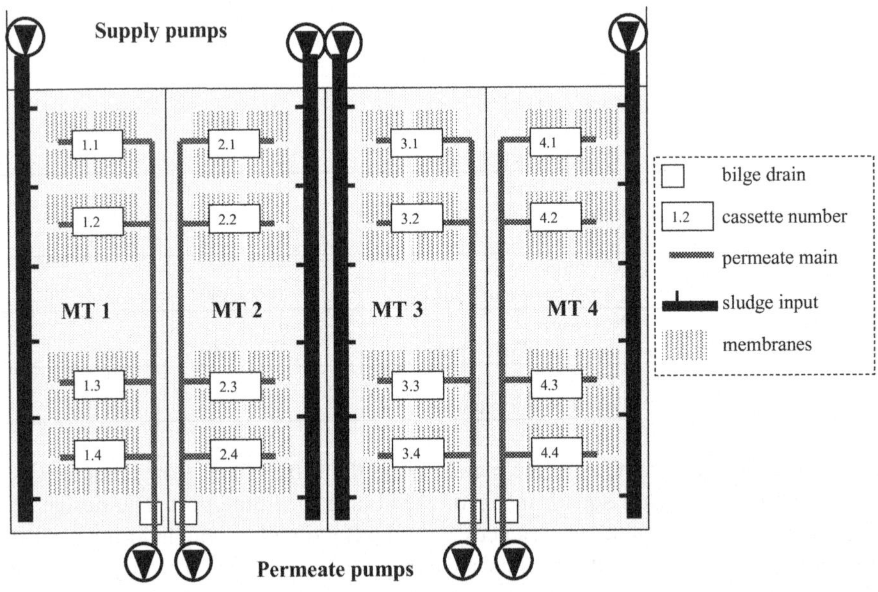

Figure 2.11. Schematic representation of the membrane tanks

Each membrane tank is equipped with a membrane influent pump where the activated sludge mixture is transported from the aeration tank. The objective of this is to refresh the sludge in the membrane tank and to prevent sludge accumulation in this tank as a result of the permeate withdrawal. The sludge recycle stream of the membrane tanks flows gravitationally to the aeration circuit. Each membrane tank has its own permeate pump used for alternate permeate withdrawal and permeate back wash. Inclusion of a back wash in the process cycle prevents the membrane surface and the membrane pores from being clogged.

The membrane modules at the bottom side are equipped with coarse aeration. The air bubbles ensure turbulence, preventing the attachment of sludge particles to the membrane surface. The aeration takes place alternatively with a switch frequency of 15 seconds. There are two types of alternating aeration possible in the system; sequential aeration and air-cycling, schematically given in Figure 2.12. During the air-cycling cassettes 1 and 3 and cassettes 2 and 4 are alternatively aerated. During sequential aeration the switch of the air input does not take place between the cassettes but in the cassettes. The advantage of air-cycling is that the turbulence in the cassettes is higher so that the operation is possibly more efficient.

In process mode the alternating aeration is continuous. The principle of the aeration in stand-by mode is that 50% of the capacity of one blower is used to aerate all membrane tanks to be aerated in the stand-by mode. All membrane streets in stand-by mode are aerated one after each other. This procedure is adjusted to be repeated for a number of times. Afterwards there is a pause when the aeration of the operating membrane streets is switched off. After this pause the aeration cycles are started again.

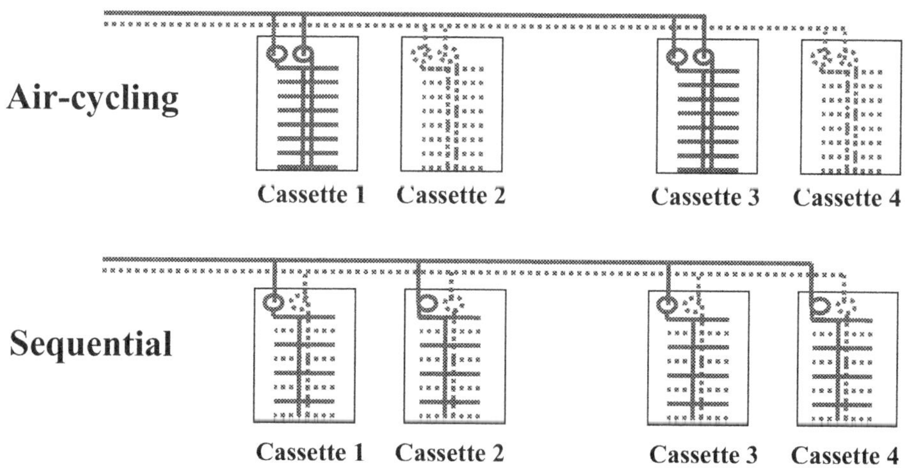

Figure 2.12. Schematic representation of air-cycling and sequential aeration.

In addition to the continuous physical cleaning of the membrane surface also periodical chemical cleaning is necessary. The idea behind this is to prevent the membranes fouling which will cause a decrease in their capacity. The cleaning method is based on so called maintenance cleanings (MC). This cleaning method was developed during the pilot-research in Beverwijk. These MCs are performed once a week or once per two weeks with significantly lower concentrated chemical solutions than those common intensive cleaning procedures applied until recently. A part of the cleaning cycle is moreover, performed with warm water thus increasing further the efficiency of the MCs.

2.5.4 Process control

The two most important states of a membrane tank are process mode and stand-by mode. In the process mode permeate pumps withdraw permeate. Here the membrane aeration and the membrane influent pumps are in operation. In stand-by mode no permeate is withdrawn. The membrane aeration is for a large part of the time off and is switched on to mix the content of the tank to prevent anaerobic conditions. The membrane influent pumps are set on a constant flow and take care of the continuous sludge refreshing.

The four membrane streets are together in process mode only when the hydraulic loading to the installation is high, thus during storm weather. During dry weather only 1 or 2 of the 4 tanks are needed to be able to handle the influent. The number of the membrane tanks in process mode is determined based on the water level in the aeration tank. When the level in the aeration tank increases (as a result of the increased hydraulic loading) a membrane tank is switched from the stand-by mode to process mode. When the water level drops a tank is switched to stand-by mode. This principle is shown in Figure 2.13 for a day with strongly fluctuating influent flow. The switching of the membrane tanks is completely automated. All membrane tanks operate in this way for the same number of working hours.

The plant operator can set in the control installation a so called optimal flux. At this optimal flux the membranes function optimally under prevailing conditions. It means that hereby almost no fouling of the membranes takes place and therefore no significant decrease of the permeability. The optimal flux determines the amount of permeate that can be withdrawn from the membrane tank. When the optimal flux is set to 20 $l/(m^2.h)$ net approximately 100 m^3/h of the permeate can be withdrawn from one membrane tank. At dry weather flow the in process mode functioning membrane operates at the optimal flux. Only the membrane tank that was switched on as the last, attempts in process mode to keep the required water level in the tank by adjusting the permeate pump. When all tanks are in process mode and the water level increases further, the withdrawal capacity is stepwise increased to the design flux of 37.5 $l/(m^2.h)$.

The permeate withdrawal cycle consists of withdrawal and back wash of permeate. Because for both functions the same pump is used, the rotation direction of this pump needs to be repeatedly changed. In practice after 6 minutes of permeate withdrawal, permeate is washed back for 20 seconds. The capacity of the wash back flow is coupled through a factor (e.g. 1.25) to the flow of the permeate withdrawal. This results in the decrease of efficiency of the filtration step, since a part of the withdrawn permeate is pumped back to the membrane tank and has again to be withdrawn. The filtration efficiency on the WWTP Varsseveld at its current set-up is approximately 80%.

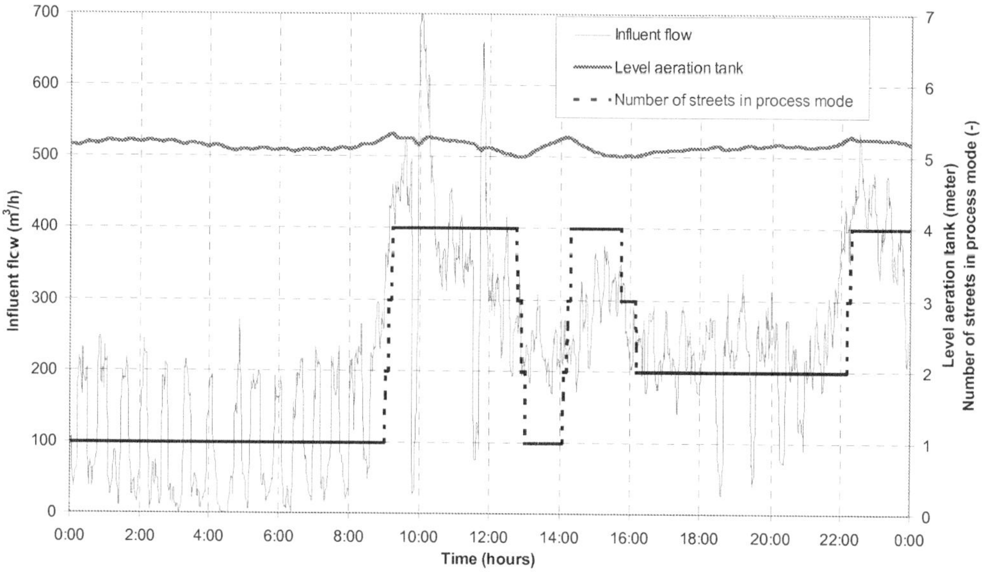

Figure 2.13. Control of membrane streets based in water level in the aeration tank

In case of a leakage in a membrane or in the permeate pipe, sludge will be withdrawn together with the permeate. Membranes would be subsequently washed back with this permeate, with as a result attachment of sludge from the outer surface of the membrane. This can lead to a (permanent) deterioration of the system performance. Leakages could be therefore detected at an early stage with turbidity meters placed in the four permeate pipes and four membrane streets. When the turbidity is above 10 NTU the concerned membrane street is switched off and the alarm is switched on.

The chemical cleaning of the membranes is performed based on a regular Maintenance Cleaning (MC). When needed an Intensive Cleaning (IC) can be performed also. The MC is applied to maintain the permeability of the membranes at a constant high level, so it is possible to work at low pressures when the membrane fouling occurs less frequently. Hereby, the employed cleaning solution concentration is relatively low, the contact time is relatively short and the frequency relatively high, for instance once or twice a week. The standard MC procedure at WWTP Varsseveld consists of the application of a strong oxidant followed by an acid. This two-step cleaning is performed with both, warm- (circa 35°C) as well as cold water. The warm water increases the efficiency of the chemicals and consequently the cleaning efficiency. An IC can be performed when the permeability decreases to an undesired low level and is focused to bring the permeability to the required level. Intensive cleanings have a low frequency but a higher chemical concentration and a longer contact time than a MC. This cleaning needs to be performed when the original permeability cannot be recovered anymore when using only the MC's. The expectation for MBR Varsseveld is that the application of MCs is efficient enough to apply the IC-procedure maximally once a year.

3

Technological results

3.1 INTRODUCTION

The MBR Varsseveld was started up on the 21st December 2004. In the first weeks the wastewater was distributed over the existing conventional WWTP and the new MBR. From the middle of January 2005 the existing system was put out of operation and the MBR was fed with the total influent. In this chapter the technological operation is described from the start-up until the end of April 2006. The description of the technological operation is divided into pre-treatment (§3.2), the biology (§3.3) and the membrane filtration (§3.4). In §3.5 the results of the study 'Specific micro-pollutants' are presented, a monitoring program where measurements were taken for the presence of hormones, pharmaceutical residues and others in the influent and the permeate of the MBR Varsseveld.

3.2 PRE-TREATMENT

3.2.1 Introduction

In the months October 2005 till January 2006 attention was paid to the sub-study "Pre-treatment" [3] on the technological operation of the pre-treatment. Hereby there was an intensive sampling program performed with the goal of obtaining more insight into the removal percentages of the three pre-treatment steps. The set-up of the pre-treatment and all internal streams having influence on the pre-treatment are schematically given in Figure 3.1.

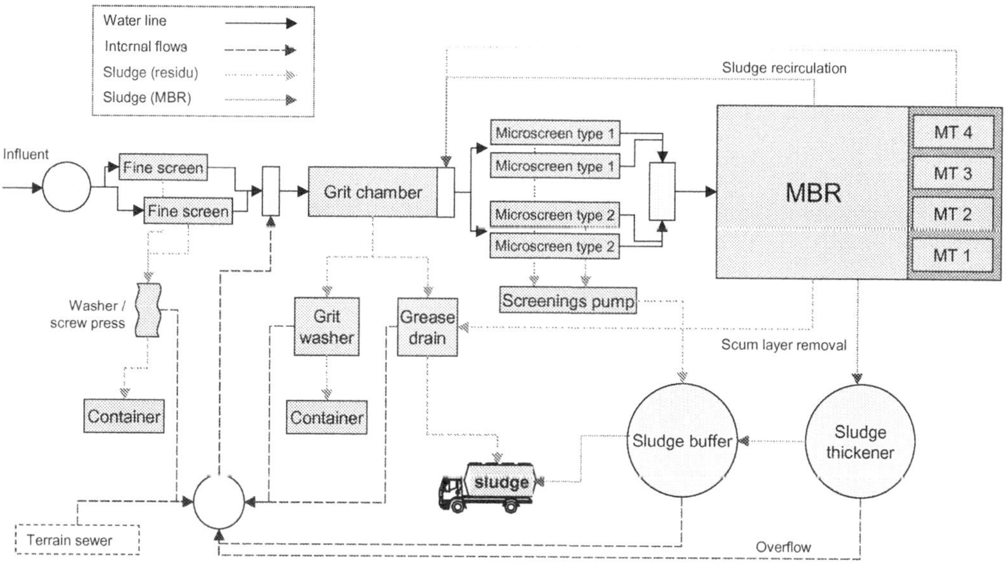

Figure 3.1. The pre-treatment and the internal streams

The results of the sub-study "Pre-treatment" are summarised in §3.2.2. The production of the different rest-streams such as screenings, micro-screen goods, grit and grease, is given in §3.2.3

3.2.2 Efficiency

To determine the removal efficiencies per pre-treatment step there was an intensive sampling- and analysis program performed. This program resulted in varying results. For some compounds the removal efficiency was not completely reliable. This was mainly caused by external disturbances of the sampled water streams resulting in unsatisfactory mass balance calculations. Sometimes there could be no explanation given for high spreading, negative removal efficiencies and not explainable mass balances.

As a consequence of these factors the efficiencies calculated per pre-treatment step do not match with the total efficiency of the whole pre-treatment (see sub-study "Pre-treatment" [3]). The results presented in Table 3.1 are therefore recalculated in the way that the total mass balance per component is closed. Hereby the influent and effluent loading is presented for each step (as percentage of the influent load set at 100%) and the removal percentage per set is presented based on the influent loading (Efficiency = - 20% means that a compound load corresponding with 20% of the influent load is removed, +7% means that the compound load increases with 7% in the considered treatment step).

Table 3.1 Efficiencies of pre-treatment (removal as % of the influent load), including the run-off water

Parameter	Fine screen			*Run-off water*	Grit/grease removal			Micro-sieves			Total
	In	Effic.	Out	*%*	In	Effic.	Out	In	Effic.	Out	Effic.
Suspended solids	100	-39	61	*+18*	79	+7	85	85	-20	65	-35
BOD	100	-26	74	*+7*	81	+11	93	93	-26	67	-33
COD	100	-34	66	*+14*	80	+9	89	89	-17	72	-28
Kjeldahl nitrogen	100	-11	89	*+8*	97	+0	97	97	-8	88	-12
Ammonium	100	-11	89	*+5*	94	-2	92	92	0	92	-8
Phosphate (total)	100	-38	62	*+13*	75	+11	86	86	+5	92	-8
Grease	100	-30	70	*+8*	78	+6	85	85	-24	60	-40

Effic., Efficiency

It is remarkable that in grit/grease chamber there was no removal of any of the influent components. Based on the measurement it seems that there was even a slight increase. Grease was reduced in the fine screens and the micro-screens. Because of the lack of representative sampling days no distinction could be made between the functioning of different types of micro-sieves.

The Technical University Kassel also performed research on the pre-treatment efficiency in months April and May 2006, with special attention to the functioning of the static sieves. Both results correspond to each other regarding the investigated components (suspended solids, COD, total nitrogen and phosphate).

3.2.3 Rest streams

The amount of the screenings was determined in two ways. Based on mass balance calculations the amount of separated suspended solids varied between 200 and 1,000 kg MLSS/day with an average of 562 kg MLSS/day. In addition the calculation of the screening production was based on the number of transported containers in the research period. Based on the containers transported, the amount of separated screen goods was approximately 11-17 kg MLSS/day. Because the difference between both methods is so enormous, their reliability is doubtful.

The amount of separated grit was registered based on the number of working hours of the grit washer. The amount of separated grit for the MBR Varsseveld was circa 2 g/p.e.day (wet volume). The amount of grease was not determined. The amount of sieve goods produced by both types of micro-sieves amounted to 200-450 kg MLSS/day. Hereby the sludge recycle from the aeration tank was regularly pumped through contributing to the relatively high amount of the sieve goods. The amount of sieve goods determined by TU Kassel amounted, without sludge recycle, for the stationary micro-sieves 175 kg MLSS/hour. Based on these data it was stated that the average sieve goods production was 200 kg MLSS/day.

3.3 BIOLOGY

3.3.1 Introduction

The biology of the MBR Varsseveld was started up from December 2004 using sludge from existing WWTP Varsseveld. The inoculum sludge was pumped from the old installation through the micro-sieves to the aeration tank of the MBR. The sludge concentration after this inoculation amounted to 2 kg MLSS/m^3. From that moment the loading of the MBR was gradually increased and the sludge concentration increased in 7 weeks to the desired level of van 10 kg MLSS/m^3. The course of the sludge concentration together with other process data is graphically shown in Figure 3.2.

During the start-up phase there was no foam- or scum layer formation. During the start-up of the simulation unit in May 2004 there was such a significant foam formation that the start-up had to be stopped (sub-study "Simulation-unit" [3]). The second start-up of the simulation unit proceeded without foaming problems, due to a gradual increase of the sludge loading. Thanks to this information the start-up of the MBR Varsseveld also went smoothly.

The most important indicators of the biology efficiency are nitrogen and phosphate removal. The nitrogen removal was optimised directly from the start-up. The optimisation of the phosphate removal was spread over the whole year. In the first months the iron salt was not dosed to see what was the efficiency of biological phosphate removal was. From June 2005 the relatively low dosing of iron salt was started; the ferric dosing pump was set to a constant flow. In the following months this flow was increased in small steps. From October 2005 the control of iron dosage was based on the wastewater flow. This adjustment resulted in effluent concentrations of 0.3-0.5 mg P/l being reached. Then, the control of the ferric dosing pump was based on a phosphate set-point of 0.15 mg PO_4-P/l.

Regarding the biological operation five periods can be distinguished as presented in Table 3.2. In presentation of the results of the biology in the first phase, the start-up is not taken into account. In this phase the biology was not functioning sufficiently and moreover the measurements from the control unit were not available. Therefore when referring to the total research period, phase 2 till 5 are considered.

Table 3.2. Research periods

Phase	Period		Days	Ferric dosing		Description / control strategy
			[-]	[l/h]	[mol Fe/mol $P_{influent}$]	
1	21/12 /2004 -	16/02/2005	58	0	0	Start-up biology
2	17/02/2004 -	31/05/2005	103	0	0	Biological P-removal
3	01/06/2005 -	30/09/2005	121	7.4	0.40	Dosing with a constant flow
4	01/10/2005 -	23/01/2006	114	17.4	0.82	Dosing based on the influent flow
5	24/01/2006 -	09/05/2006	105	21.8	1.12	Dosing based on the set-point P = 0.15 mg/l
Total	17/02 /2004 -	09/05/2006	443	11.6	0.53	Phase 2 till 5

In this paragraph the most important results of the biology of the MBR Varsseveld are presented. In §3.3.2. the most important process data throughout the research period are given. In §3.3.3. and §3.3.4 the results of the nitrogen and phosphate removal are reported. Finally the sludge production is discussed in §3.3.5.

3.3.2 Process data

The most important process parameters of the biology are graphically presented in Figure 3.2. The sludge concentration of 10 kg MLSS/m^3 is maintained relatively constantly. The strong decrease of the sludge volume index (SVI) in June 2005 is remarkable. This was the result of the ferric dosing that was started at that moment. In consequence there was then also a clear increase in the ashes fraction of the sludge.

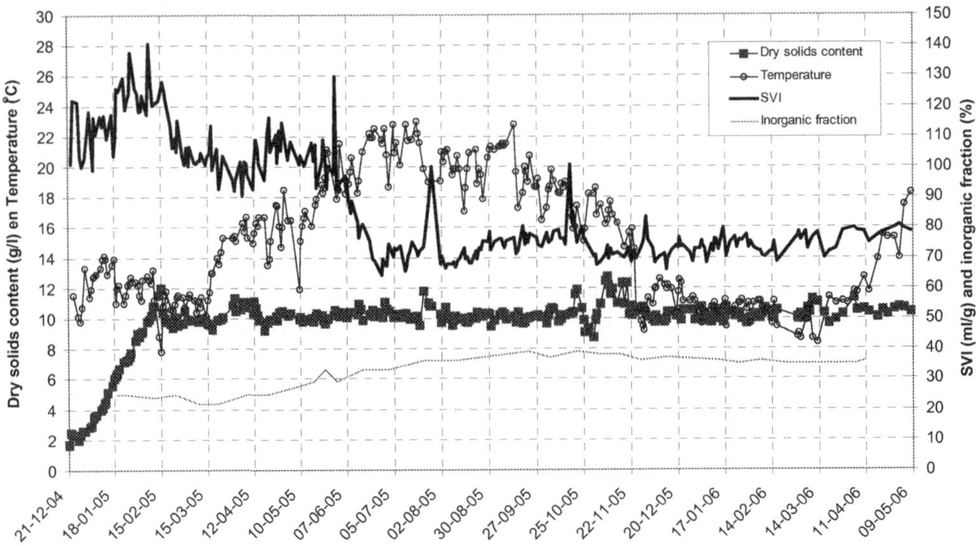

Figure 3.2. Process data of the biological treatment

The average values of the most important process parameters during the different research periods are given in Table 3.3. The sludge production is based on the average values from four different determination methods (see §5.3).

The total loading based on the pollution equivalents (at 136 g Total Oxygen Demand, TOD) is 15-20% lower than design loading. This is because of the lower nitrogen load in the influent. The sludge loading as reported in Table 3.3 is based on the influent composition after the pre-treatment Hereby the reduction in concentrations as a result of pre-treatment as presented in Table 3.1 was assumed.

For the determination of the design loading it was assumed, to be secure, that the reduction of all components during the pre-treatment was 0%. As a result of this the actual COD loading of the aeration basin is more than 20% lower than according to the design (100 g COD/(kg MLSS.d)). The average nitrogen loading was even more than 30% lower than that of the design (9.3 g N/(kg MLSS.d)). The phosphate loading was in contrast almost 20% higher than the design loading (1.66 g P/(kg MLSS.d)).

The average influent- and permeate concentrations of COD, nitrogen and phosphate during the subsequent research phases are given in Table 3.4. The COD removal during the entire period amounted to 97%, whereby the permeate concentration was very stable (22 ± 3.1 mg/l). During the storm weather the COD of the permeate dropped to a minimal concentration of 10 mg/l.

The results of the nitrogen- and phosphate removal are further discussed in §3.3.3 and 3.3.4.

Table 3.3.The most important process parameters of the MBR Varsseveld

Parameter	unit	Total	Phase 2	Phase 3	Phase 4	Phase 5
Ferric dosing	*mol Fe/mol $P_{influent}$*	*0.53*	*0*	*0.40*	*0.82*	*1.12*
Loading	p.e. (a 54 g BOD)	22,600	22,200	23,800	22,600	20,600
	p.e. (a 136 g TOD)	25,800	25,400	27,200	25,800	23,600
Sludge loading[1)]	g COD/(kg MLSS.d)	77	76	81	77	70
	g N/(kg MLSS.d)	6.3	6.3	6.5	6.1	5.9
	g P/(kg MLSS.d)	1.95	1.82	2.10	2.01	1.76
Flow	m^3/d	4,180	3,910	4,090	4,500	4,320
Process temperature	°C	15.6	14.6	20.3	13.9	11.5
Sludge content	kg MLSS/m^3	10.3	10.1	10.2	10.4	10.3
Inorganic fraction	%	33	26	35	37	35
SVI	ml/g	82	104	75	74	76
Sludge production[2)]	kg MLSS/d	1,130	1,090	1,070	1,210	1,150
Sludge age	d	34	34	35	32	33

[1)] calculated based on the volume of the aeration tank (excluding the membrane tanks)
[2)] based on the average sludge production as calculated in Table 5.1

Table 3.4. Influent (raw) and permeate concentration (calculated average)

Parameter		Unit	Total	Phase 2	Phase 3	Phase 4	Phase 5
Ferric dosing		*mol Fe/mol $P_{influent}$*	*0.53*	*0*	*0.40*	*0.82*	*1.12*
COD	Influent	mg/l	790	830	850	730	700
	Permeate	mg/l	22	25	22	20	20
	Reduction	%	97	97	97	97	97
N_{Kj}	Influent	mg/l	52	56	55	48	47
	Permeate	mg/l	1.0	1.1	0.9	0.9	1.2
NH_4-N	Influent	mg/l	30	25	30	34	30
	Permeate	mg/l	0.26	0.24	0.14	0.32	0.44
NO_3-N	Influent	mg/l	1.9	4.2	1.0	1.7	1.8
	Permeate	mg/l	2.7	2.2	2.4	3.4	2.5
N_{total}	Influent	mg/l	54	60	56	49	49
	Permeate	mg/l	3.7	3.2	3.3	4.4	3.7
	Reduction	%	93	95	94	91	92
P_{total}	Influent	mg/l	15.6	15.5	17.1	14.9	13.7
	Permeate	mg/l	2.1	4.9	2.4	0.44	0.19
	Reduction	%	87	68	86	97	99

3.3.3. Nitrogen removal

The course of the nitrogen content in the permeate is shown in Figure 3.3. The average influent- and permeate concentrations are presented in Table 3.4. The target value of 5 mg N_{total}/l required in the summer months were easily reached. In the winter of 2005 the effluent concentration was a few times above 5 mg/l, mainly caused by technical disturbance or technological operational problems. The examples explaining both situations are given below:

End October 2005 a thick scum layer occurred in the circuit system; the floating on-line analysers could not function anymore. It resulted in poor functioning aeration control and a high nitrate peak in the permeate.

End November 2005 the internal recycle pumps, the stirrers in the pre-denitrification tank and the impellers in the circuit were out of operation for almost 2 weeks. This led to a relatively high nitrate peak around 22 November 2005.

The ammonium concentration in the permeate was always low thanks to the excessive aeration in the membrane tank. In the course of time the ammonium concentration increased from 0.1-0.2 mg NH_4-N/l in phases 2 and 3, and to 0.3-0.4 mg NH_4-N/l in phases 4 and 5. This was the result of the nitrogen set-point increase in the aeration control (see §4.3.2).

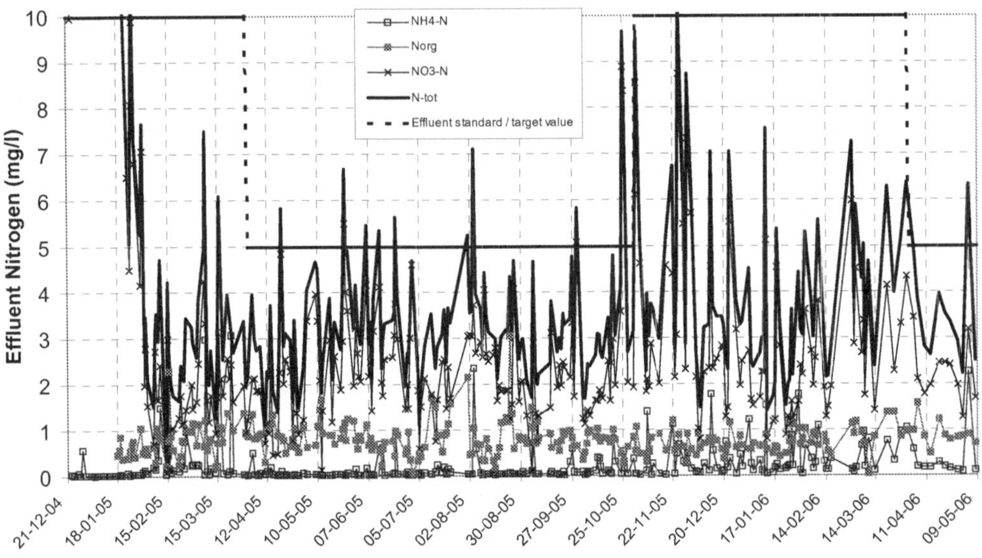

Figure 3.3 The course of the nitrogen concentration in the permeate

3.3.4 Phosphate removal

The wastewater of the WWTP Varsseveld has a relatively high phosphate content. The average phosphate concentration in the influent in the year 2005 was approximately 15 mg P_{total}/l. The target effluent value for the MBR is 0.15 mg P_{total}/l, which means that a removal efficiency of 99% has to be reached. The results of the phosphate removal for different research periods are given in Table 3.4. The course of the phosphate concentration in the influent and the permeate and the specific ferric dosing, expressed in mol iron per mol phosphate in the influent, are given in Figure 3.4.

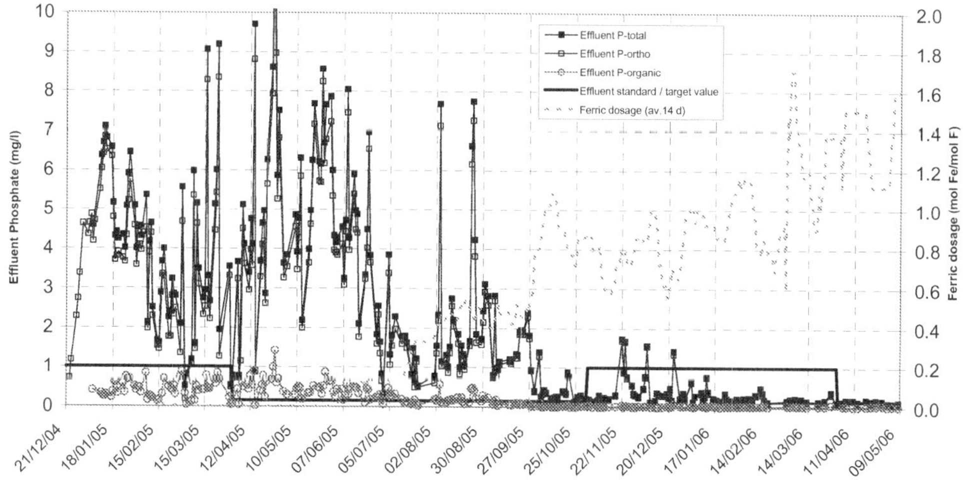

Figure 3.4. The course of the phosphate content in the permeate and the specific iron dosage

Without ferric dosing a permeate concentration of 4-5 mg P_{total}/l was reached. The target value of 0.15 mg P_{total}/l was reached in the last part of phase 5. In Figure 3.5 the course of the phosphate concentration in this period is shown in a detailed way. The necessary ferric dosing hereby was 1.2-1.4 mol Fe/mol $P_{influent}$.

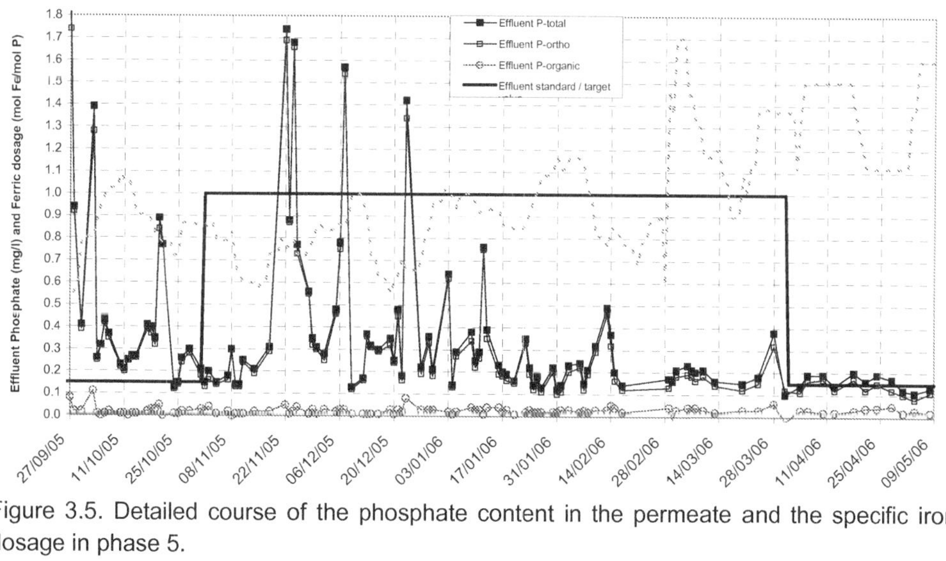

Figure 3.5. Detailed course of the phosphate content in the permeate and the specific iron dosage in phase 5.

3.3.5 Sludge production

The Water Board registers carefully how much sludge is transported from the installation. The transported sludge consists of the excess sludge from the aeration tank and, the screenings of the micro-sieves and the content of the scum layer/grease container (see Figure 3.1). Of this stream the volume (cubic meters) and the MLSS content were measured. The MLSS content of the sludge is determined in grab samples. Based on this the daily withdrawal rate was calculated for different research periods as presented in Table 3.5. The results are worked out in §5.3.

In the framework of the sub-study "Pre-treatment" [3] it was attempted to quantify the amount of produced micro-sieve goods and grease. For both streams it was quite difficult. Therefore a global estimation was done for both rest streams. For the calculation of the amounts of sieve goods the default values were used [5]. The amount of grease is probably negligible in relation to remaining streams and therefore set to zero in these calculations.

Table 3.5. Sludge production (including sieve goods) of the WWTP Varsseveld based on the transport data

Parameter	**Unit**	**Conventional**[1)]	**Phase 2-5**	**Phase 2**	**Phase 3**	**Phase 4**	**Phase 5**
Ferric dosing	*mol Fe/mol P_{infl}*	*0.5*	*0.53*	*0*	*0.40*	*0.82*	*1.12*
Transport	kg MLSS/day	1,320	1,230	1,290	1,010	1,370	1,090
Fraction sieve goods	kg MLSS/day	-	200	200	200	200	200
Fraction grease	kg MLSS/day	-	0	0	0	0	0
Excess sludge (incl.scum layer)	kg MLSS/day	1,320	1,030	1,090	810	1,170	890

1) The sludge transport of the old (high loaded) conventional WWTP Varsseveld in the year 2004

3.3.6 α-factor

As a result of the high sludge concentration in the aeration tank the viscosity of the activated sludge increases. Due to this the oxygen transfer and the α-factor worsen. During the pilot-research in Beverwijk the α-factor varied between 0.4 and 0.7 [1,2]. In the design of the MBR Varsseveld an α-factor of 0.5 was assumed.

In the framework of the sub-research "OC and hydraulics" [3] in the year 2005 the α-factor was measured a few times in the experimental set-up. The measurements are graphically shown in Figure 3.6. During the start-up phase the α-factor was relatively high as a result of the lower sludge concentration at that time. After the sludge concentration reached a desired level, the α-factor decreased to the design level. In the course of the year the α-factor increased to a level of 0.7-0.8.

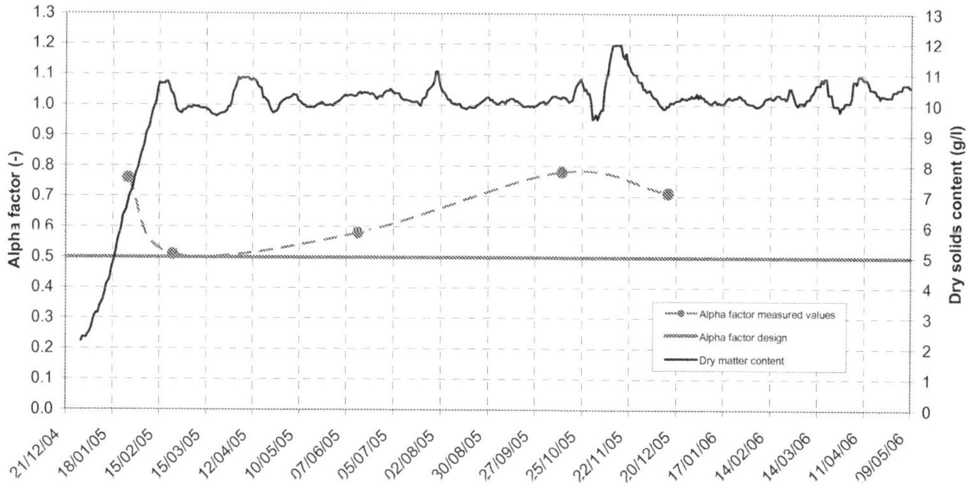

Figure 3.6. The course of the α-factor in time

3.4 MEMBRANE FILTRATION

3.4.1 Introduction

The membrane installation was from 21st December 2004 in operation, just like the biological part. During the first weeks, the influent flow was relatively low, since only a part of the wastewater was treated in the MBR. An overview of the loading of the four membrane streets throughout the whole research period (phase 1 till 5) is given in Table 3.6. From this overview one can see that the membranes were for 45% of the time in operation. The hydraulic loading of all four streets was nearly similar. The performance of the membranes is described in §3..4.2 The results of the guarantee tests are given in §3.4.3.

Table 3.6 Important process data of the four membrane streets (21st December 2004 – 9th May 2006)

Parameter	Unit	MT1	MT2	MT3	MT4	Total
Total flow	m^3	531,700	544,100	552,700	533,100	2,161,600
Average flow	m^3/d	1,055	1,080	1,097	1,058	4,289
Average flux (net)	$l/(m^2.h)$	8.7	8.9	9.1	8.7	8.9
Specific flow	m^3/m^2	105	108	110	106	107
Time in process mode	%	45	45	45	44	45

3.4.2. Membrane performance

As it can be seen from Table 3.6 each membrane street produced in 16 months time more than 500,000 m^3 permeate. Each street operated for 45% of the time in process mode and 55% in a standby mode. The average net flux, spread over the whole period, amounted to approximately 9 l/(m^2.h), corresponding to a net flux during process mode of circa 20 l/(m^2.h). The course of the net flux, the permeability and the process temperature for one of the four membrane tanks is shown in Figure 3.7.

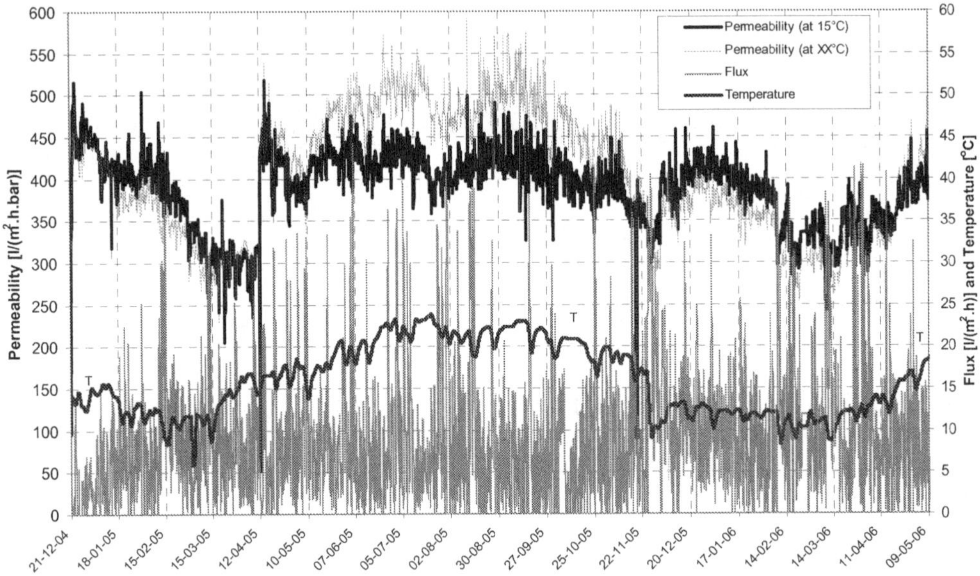

Figure 3.7 The course of the flux, permeability and temperature of one of the membrane tanks

A notable period was in months February and March 2005. In this period, directly after the start-up of the MBR Varsseveld, the permeability decreased quite rapidly. The reason for this was the presence of an industrial polymer (PVA) in the wastewater, causing sticking of the membranes (see §4.4). After an intensive cleaning of the membranes in April 2005 and the separate discharge of this influent component, the permeability increased to the initial level. From that time the permeability remained relatively stable between 300 - 400 l/(m^2.h.bar). In the winter of 2005 there was a decrease of the permeability as a result of the low temperature; an aspect further discussed in §5.8. Until November 2005 chemical cleaning of the membranes (MC) was performed each 2 to 3 weeks. This frequency was increased in the winter to once per week. Also the cleaning procedure was adjusted at that time (see §4.4.2).

3.4.3 Guarantee tests

The design flux of the MBR Varsseveld was relatively high when compared with MBR-installations in other countries (§2.5.2). The net design flux was 37.5 l/(m^2.h) at a storm weather flow of maximally 3 days. In case one of the four membrane streets was out of operation, the installation was meant to function at a maximum net flux of 50 l/(m^2.h) for a period of 8 hours.

In order to obtain the guarantee of the membrane supplier the initial design conditions had to be tested. This guarantee test was performed in the beginning of October 2005 in membrane tank 3; the membrane tank that had the lowest permeability. The results of these guarantee measurements are shown in Figure 3.8.

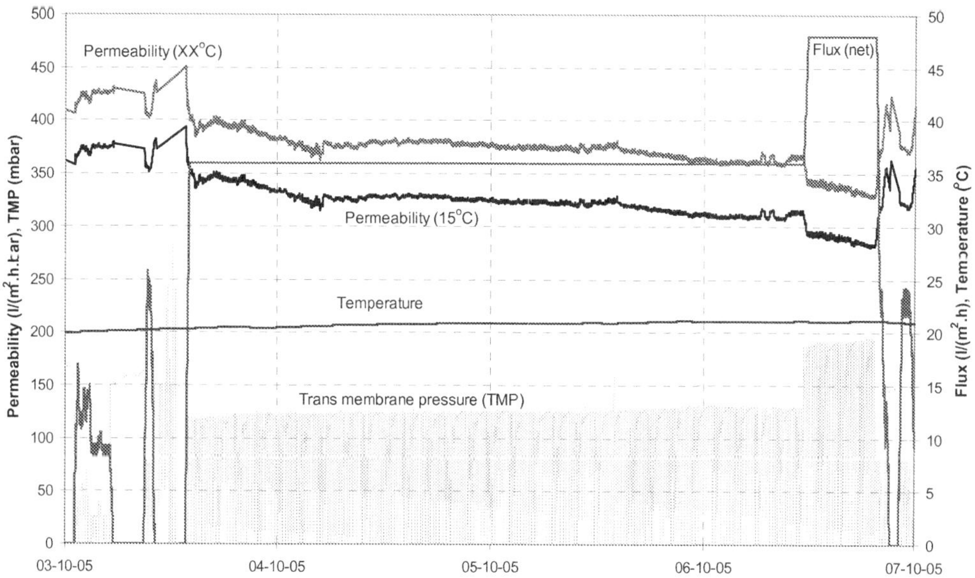

Figure 3.8. The results of the guarantee test (membrane street 3)

Although the net flux was not completely at the desired level, it can be concluded from the result of the test that the membranes are able to maintain the design flux for the length of the demanded period. The permeability decreased to circa 360 l/(m^2.h.bar) at optimal flux to circa 310 l/(m^2.h.bar) after almost 3 days of design flux. The 8 hour maximum flux resulted in a following decrease of circa 280 l/(m^2.h.bar). After the guarantee test the permeability at optimal flux increased to 330 l/(m^2.h.bar). Following maintenance cleaning on the 18th of October the permeability was again recovered to the level of the period before the guarantee meetings. Based on the results it was concluded that the membranes could provide the guaranteed performance. It has to be added at this point that the process temperature was relatively high (21°C).

Although the guarantee tests were not repeated in the winter it is possible to evaluate the performance of the membrane at low temperatures and high fluxes. On the 8th and 9th February 2006 there was a long lasting rain period, whereby the membranes operated for almost 24 hours at the maximal flux and a process temperature that decreased from 12.4 to 7.7°C within 12 hours. The results of this event are given in Figure 3.9.

The operation of the membranes during a relatively short lasting rain weather flow in the winter is obviously worse than at a long lasting peak loading in the summer. The average net flux amounted to approximately 32 l/(m^2.h), which was lower than the design flux. The permeability dropped in 20 hours from circa 350 l/(m^2.h.bar) to circa 255 l/(m^2.h.bar). The trans-membrane pressure increased thereby to circa 200 mbar. In the following 2 months the permeability remained relatively constant between 300 and 350 l/(m^2.h.bar), also some heavy rains occurring during this period (Figure 3.7).

Although the guarantee tests proved that (in the summer) the membranes can handle the design flux and the maximum flux for a period of 3 days and 8 hours respectively, the same could not be completely proved in a winter situation.

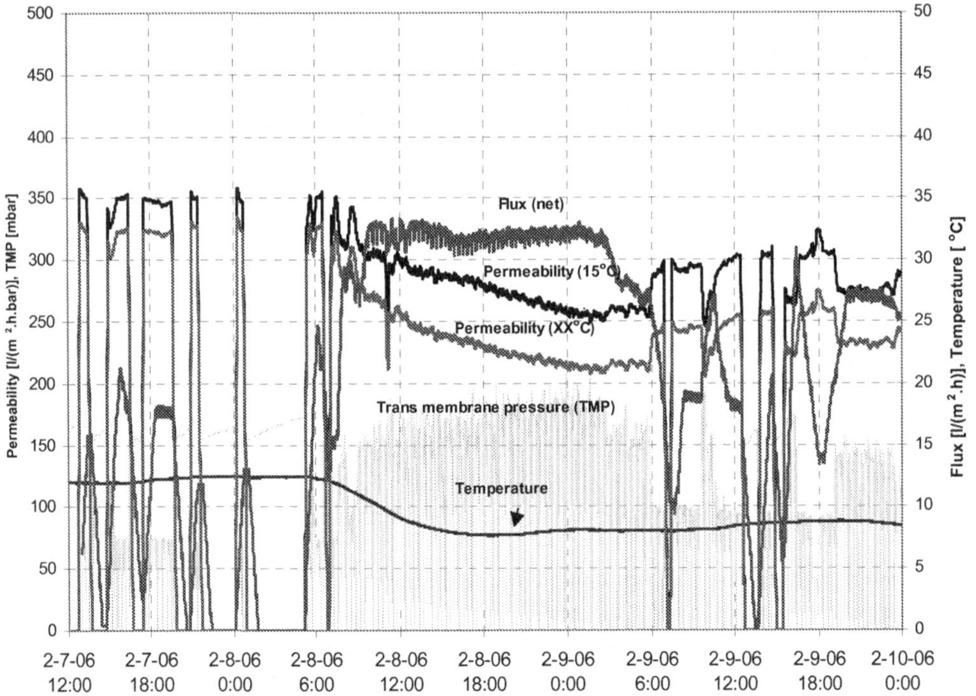

Figure 3.9 The performance of the membranes at low temperature and high flux

3.5 MICROPOLLUTANTS

3.5.1 Introduction

The European Water Directive and the new European guideline for safe recreational waters will possible require the highly efficient treatment of wastewater. The MBR is one of the possible techniques. In the framework of the research at the MBR Varsseveld there was, therefore, special attention paid to the specific pollutants in the effluent such as endocrine disrupting compounds, pharmaceutical residues and pathogens. In this section the results of this research are summarised. The set-up and the detailed results of this research are described in the sub-research report "Specific micropollutants" [3].

3.5.2 Monitoring program

The monitoring program consisted of two measurements in influent and four measurements in effluent for the general parameters, PAH's, phtalates, endocrine disrupting activity, pharmaceuticals, screening for unknown compounds and bacteriological parameters. In addition, as a part of a so called "Total Effluent Assessment", the acute toxicity for bacteria, algae and water fleas was determined as well as the bioaccumulation (the accumulation of the compounds in organisms).

In the research the results of regular measurements by the Water Board were also included. This included heavy metals, EOX, organic chlorine, N- and P-pesticides and PCB's.

There was also the EOX-content measured twice in the effluent after the membrane compartment was cleaned. The results of these measurements are presented in §5.8.

3.5.3 Results and conclusions

In the influent of the MBR Varsseveld heavy metals, PAH's, phtalates, endocrine disrupting compounds, pharmaceuticals, toxicity and potential bioaccumulating compounds were found. PCB's and pesticides were not detected in the influent.

The compounds associated with the suspended material, such as phtalates and higher PACs were removed to a high extend as expected. Also almost all other compounds were removed to a satisfactory degree. In the effluent some PACs, metals (except for cadmium and mercury), endocrine disrupters, pharmaceuticals, and unknown compounds were still found. Incidentally the pesticide diazinon was found in the effluent in a concentration of maximally 0.02 μg/l.

The estrogenic activity in the effluent, measured with the ER-CALUX assay, was negligible, lower than <0.5 ng EEQ/l. The pharmaceutical concentrations were below the detection limit, and when the compounds were detected they were in the range of tens of μg/l. An exception is the β-blocker metoprolol. This often consumed agent was found in the effluent at concentrations of more than 2 μg/l.

The effluent after concentration was toxic for bacteria, algae and water fleas. The measured toxicity in the Microtox test could be considered as normal, while the toxicity of the effluent for algae seemed to be high when compared to conventional systems. The number of measurements was however quite limited.

There were some potentially bio-accumulative compounds detected after extraction with a SPME-fibre. This extraction is supposed to simulate the bio-accumulation of soluble, biologically available compounds in organisms. The measured concentrations are a few times higher than was found in earlier research performed at a limited number of WWTPs.

There were no *E. coli* bacteria and intestinal enterococus detected in the effluent. From this point of view the effluent complied with the standards for the new European Swimming

Water Directive. Also F-specific bacteriophages, as indicators for the presence of viruses, could not be detected.

The concentrations of nickel and lead in the effluent exceed the proposed standards of the Water Framework Directive for the surface water. The magnitude of the excess is however such that it can be expected that the contribution of the effluent of the MBR will not lead to exceeding of the standard in surface water. The concentrations of cadmium, mercury and Di (2-ethylhexyl) phthalate (DEHP) complied with the current WFD standards. The remaining priority compounds (some PAH's and pesticides) were not detected in the effluent.

4

Operation

4.1 INTRODUCTION

In comparison with a conventional activated sludge system an MBR has a number of process components, for which there is little experience in the Netherlands. The membranes and all associated parts are certainly the most important elements. But also the intensive pre-treatment with use of micro-sieves, is new. The aeration tank is very compact, has a high sludge content and a separated floating layer withdrawal system. These unknown process units and process conditions lead to new operational experiences and problems. In this section an overview is given of the most important operational and technical aspects investigated during the research period. Hereby there is distinction made for the pre-treatment (§4.2), the biology (§4.3) and the membrane filtration (§4.4).

4.2 PRE-TREATMENT

4.2.1 Introduction

During the first 1.5 year that the MBR Varsseveld was in operation, major attention was paid to the performance of the pre-treatment. Both the fine screens and the micro-sieves were designed with 100% redundancy. Because there was no by-pass of the fine-screens and micro-sieves, during a process disturbance of one of these units the influent to WWTP was stopped. The urgency alarm then went on while the wastewater was buffered in the sewer. This situation occurred several times and led to the introduction of a few correction measures in the process operation. The functioning of the micro-sieves turned out to be the most critical. These are, in turn, dependent on the good operation of the preceding fine screens and grit/grease removing installation. Optimisation focussed therefore on the pre-treatment as a whole. The most important measures and experiences with the fine screens, grit/grease removing and micro-sieves are subsequently described in §4.2.2. and §4.2.3.

4.2.2 Fine screens

The fine screens operated for the whole research period reliably and without any disturbance. As a consequence of the operational problems with the micro-sieves (see §4.2.4) however there were introduced some adjustments in the control of the fine-screens.

The process control of the fine screens was in the first instance the same as in many conventional WWTPs, which means that after reaching a switch on level the screen step moved one step forward. When the water level before the screen increased again, one waited until the high level was reached, and the screen was then completely cleaned. It occurred regularly in practice because the influent fluctuations at the WWTP Varsseveld are significant. As a result the fine screen was cleaned completely to make the bar opening of 6 mm completely available again. In such a situation the filtering capacity of the screen is not present anymore, so a large part of the suspended solids will reach the micro-sieves. This could lead to blockage of the micro-sieves, especially at higher influent flows.

In the beginning of 2006 the control of the fine screens was adjusted in such a way that when the fine screen reached the switch level it did not make one cleaning step, but many until the water level was again below the switching level. As a result the fine screen was covered for a longer period with a layer of screenings and the micro-sieves received smaller influent loadings and did not become clogged so fast.

Although the performance of the fine screens improved, it did not avoid the micro-sieves being overloaded from time to time. Increase in flow rates higher than 300 m^3/h within a time interval of 2 minutes led to such a water level increase before the fine screen, that still a high level was reached and the screen was completely cleaned.

Therefore it was decided to perform an experiment with two fine screens working in parallel. Due to the doubling of the capacity the water level before the fine screens did not increase so quickly and the mentioned cleaning philosophy would function better. Since the 1st March 2006 both screening removing streets operate in parallel.

4.2.3 Grit/grease removal

During the first months after the start-up the grit/grease removal caused odour nuisance for the WWTP and surrounding area. After some adjustments (e.g. increase of the ventilation capacity, equal air distribution) this problem was solved. Furthermore, the grit/grease removal did not cause any technical or technological problems. Based on a visual evaluation of the turbulences in the grit and grease chamber the air input was lowered in the grit chamber. The installation was never put out of operation because of technical nuisances.

4.2.4 Micro-sieves

The micro-sieves were considered as the most sensitive units of the MBR Varsseveld. In the first 1.5 years of operation there were many operational disturbances, especially during fast changing influent loading at the beginning, and also at the end due to the storm weather flow. In these situations high loadings of non-dissolved material and possibly grease from the sewer were transported to the micro-sieves causing their heavy overloading, despite the presence of fine-screens and grit/grease chambers. This resulted in the discharge of the screenings from the micro-sieves being temporarily not sufficient causing accumulation of wastewater in the machines and/or too much wastewater being discharged together with the screenings. Under these circumstances a high level in the micro-sieves or a high level in the transport were signalised and the micro-sieve street was switched off.

In the first 1.5 year after the start-up of the installation there were some measures taken to improve the operation of the micro-sieves. In addition to the already mentioned adjustments in the operation and process control of the fine screens and the grit chamber, the operation of the micro-sieves was adapted. The most important adjustment comprised putting in operation all four machines during a high influent flow. Hereby the full reserve capacity was used under normal process conditions. In addition the capacity of the sieve goods transport pumps was increased. There were also some little adjustments performed in the sieves themselves.

Type 1 is a compact rotating drum sieve (manufacturer Auxill) where the sieve goods are removed using sprinkle installation. These micro sieves are delivered in the first instance with fine meshed gauze stringing. Before putting into operation, this stringing is replaced by a perforated plate which does not have any consequence for the size of the drum. Because the water permeability surface of the perforated plate is smaller than that of the gauze stringing, problems occurred with the transport of the sieve goods from the machine. Mounting the sprinkle installation in the discharge drain solved the problem. Recently it was concluded that the use of the perforated plate led to decreased capacity of the machine and it was decided to replace them with larger types. Based on these operational experiences, the machine has been further optimised (improved seals, sieve goods transport with screw).

Type 2 is a stationary sieve (manufacturer Contec) whereby the sieve goods are removed from the sieve surface using brushes. During a control of membranes in February 2006 there was brush hair found in the membranes from these micro sieves (§4.4.5). There were also fibres and hair found. The reason for this was a leakage in the seal of the stationary sieves. Directly after installation of these sieves there were tests performed with dosing of hair in the influent of the micro sieves. It was then concluded that the micro-sieves did not let the hair through. The brush hair and other material reaching the membranes probably got through the margin in the overlapping between the sieve plates. The micro sieve consists of two sieve plates overlapping with each other. This plate overlap was additionally strengthened, thus not allowing hair and fibres to pass through. In addition there were a number of measures taken to improve the working of the brushes and to minimise the water discharge together with the brush hair. This comprised the cutting of a part of the brush hair, mounting brushes closer on the sieve plates and replacing the brushes with another type.

After all these optimisations in the period between 1st March 2006 till 1st June 2006 the high water in the stationary sieves occurred just once. After the automatic change to the drum sieves, the micro sieves could function without any limitations regarding the influent. It is expected that after the change of the drum sieves to the sieves of a larger capacity, the combination of both micro sieves will have a sufficient capacity to handle the incidental loading peaks without any problems. As a precaution, once a week the influent pumps operate for a short time at their maximal capacity to prevent the accumulation of material in the influent main.

4.3 BIOLOGY

4.3.1 Introduction

The biological treatment performed in general very well during the research period. The effluent requirements for nitrogen were easily attained and the phosphorus removal reached, according to the planning, the required level. The aspects deserving the most attention are discussed in this section. These are subsequently the aeration control (§4.3.2), the ferric dosing (§4.3.3) and the scum layer forming (§4.3.4).

4.3.2 Aeration control

The aeration was controlled using a cascade control of nitrogen and oxygen. The nitrogen set-point was defined as a ration between the ammonium content and the sum of the ammonium- and nitrate content in the aeration tank (Figure 2.6). Hereby on-line ammonium and nitrate analysers were used in the aeration tank. The concentrations of ammonium and nitrate in the permeate were not the same. Because the content of the membrane tank is aerated a part of the ammonium will be oxidised to nitrate. The $NH_4/(NH_4+NO_3)$ ratio in the aeration tank is therefore higher than in the permeate.

In the framework of the sub-research "SIMBA-modelling" [3] there was a relation established using model calculations between the nitrogen set-point and the permeate quality. The results of these calculations are summarised in Table 4.1. The calculations are performed at an average dry weather flow, a constant ferric dosing of 20 l/h and a process temperature of 12°C.

Table 4.1.Results of simulation calculations of the aeration control

Set-point N-ratio[1)]	Aeration tank				Permeate			
	NH_4-N	NO_3-N	NH_4-N + NO_3-N	N-ratio[1)]	NH_4-N	NO_3-N	NH_4-N + NO_3-N	N-ratio[1)]
[-]	[mg/l]	[mg/l]	[mg/l]	[-]	[mg/l]	[mg/l]	[mg/l]	[-]
0.1	0.23	2.32	2.55	0.09	0.08	2.47	2.55	0.03
0.2	0.38	1.59	1.97	0.19	0.12	1.89	2.01	0.06
0.3	0.54	1.28	1.82	0.30	0.17	1.66	1.83	0.09
0.4	0.68	1.05	1.73	0.39	0.22	1.49	1.71	0.13
0.5	0.81	0.90	1.71	0.47	0.26	1.42	1.68	0.15

[1)] the N-ratio is calculated as NH_4-N/ (NH_4-N + NO_3-N)

From the results it can be seen that the N-ratio in the permeate is on average 3 times lower than in the aeration tank. Setting the nitrogen set-point too low leads to unnecessarily low ammonium concentrations in the permeate and too excessive aeration. The elevation of the nitrogen set-point from 0.1 to 0.5 results, based on the model calculations, in an increase of the average ammonium concentration in the permeate from 0.08 tot 0.26 mg NH_4-N/l. The nitrate concentration decreases hereby from 2.47 to 1.42 mg NO_3-N/l. Too high a set-point is not desirable because of the risk that the aeration will not be sufficient to ensure good sludge properties. In practice from the mid 2005 a nitrogen set-point of 0.3 was applied. A further increase of this set-point leads to a relatively insignificant improvement of the effluent quality.

4.3.3 Ferric dosing

The ferric dosing for phosphate removal always took place in the aeration tank (Figure 2.6). In this point the phosphate content is relatively low because of the biological uptake. Possibly due to this the efficiency of dosing is lower and a larger dosing capacity is needed. An alternative dosing point is the pre-denitrification tank. This functions also as an anaerobic tank where phosphate release takes place. An ferric dosing in the influent to the pre-denitrification tank can possibly lead to a lower dosage flow. There is a risk however, that the biological phosphate removal will be disturbed. Using model calculations the effect of the dosing location on biological and chemical phosphate removal was investigated (sub-research "SIMBA-modelling" [3]). From this study it turned out that there was no difference between dosing in the aeration tank and in the pre-denitrification tank. It has to be mentioned however that there were some doubts regarding the reliability of the chemical part of the applied phosphate removal model.

The location of the iron dosage was moved in March 2006 from the point indicated in Figure 2.6 location (downstream of the first aeration field) to the point at the outflow to the membrane tanks. The reason was the increase in the inorganic contamination of the membranes stated during the optimisation research of the membrane cleaning (sub-research "Membranes", [3]). Because of this the ferric dosing was moved to the location where the mixing was the most turbulent and the distance from the membranes was the most distinct

The dosing of the iron salt can be manually controlled, based on the influent flow and phosphate content. As shown in Table 3.2 all three control strategies were implemented in the research period. With the control based on the phosphate content in the aeration tank the phosphate concentration in the permeate was significantly lowered and in general the set-point was reached. Because the maximum capacity of the dosing pump was limited, the phosphate content at peak influent flow was sometimes above the set-point.

4.3.4 Floating layer

After the start-up of the installation there was sometimes a scum layer present of 10-20 cm in thickness. The scum layer accumulated namely in the curves of the aeration circuit where it could not pass the aeration packages. The scraper of the scum layer worked satisfactory but it was not able to prevent the scum layer completely. The scum scraper was also only operational for a part of the time because the water level in the aeration tank varied. The aeration tank is namely used to flatten the fluctuations in the influent, in order to secure the influent to the membrane installation as constantly as possible.

In the course of the research period, the significant scum layer was formed three times, in February, July and October 2005. These scum layers led to operational problems, amongst others, the floating analysers could not function well (Figure 4.1) and the aeration became disordered. It can be seen in Figure 3.3 that two weeks after the scum layer occurred, 17th

October 2005, the nitrogen removal significantly deteriorated, because the membrane of the oxygen analyser dried out, the analyser and therefore also the aeration control did not function properly. In July and October 2005 the scum layer was so thick (> 1 meter) that it flew over the edge of the aeration tank.

Figure 4.1. Floating scum layer in the aeration tank (left) and the sprinkle nozzles (right)

To prevent the problems with scum layers a permanent sprinkle installation was installed at the end of October 2005. The effect of this sprinkle set-up on the process water use is given in Figure 4.2. In the beginning a lot of sprinkle water was used because the sprinkle nozzles could not be well adjusted (Figure 4.1 right). In the course of 2006 the flow through the nozzles was lowered. As a result of the scum layer sprinkling, the average consumption of process water increased from circa 200 to 350 m^3/d. This leads to an additional loading of the membranes of 150 m^3/d, corresponding with circa 4% of the influent flow.

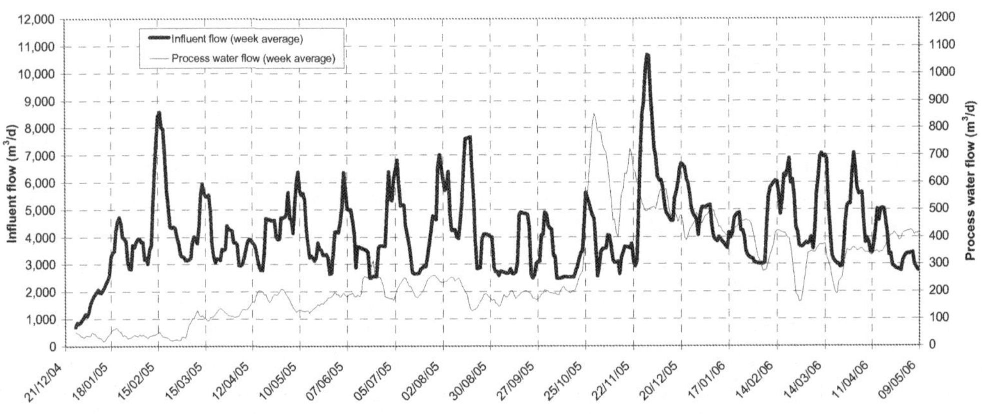

Figure 4.2. The process water consumption on the MBR Varsseveld

4.4 MEMBRANE FILTRATION

4.4.1 Introduction

In the first year after the start-up of the MBR Varsseveld much attention was paid to the optimisation of the process set-ups of the membrane filtration installation. An extended description of this is to be found in the sub-study "Membranes" [3]. In this section the most important aspects in relation to the operation are discussed. The presence of an industrial polymer in the wastewater originating from a local industry caused in the first months after the start-up large operational problems. The influence on the performance and the solution of the problem are described in 4.4.3. A second problem associated with installation was of a technical nature and caused leakages in the membrane modules (§4.4.4). A third aspect which deserved attention was the macro-fouling observed in the membrane cassettes (§4.4.5). To start with an overview of the most important optimisations of the membrane installation is given in §4.4.2.

4.4.2 Optimisation of process operation

In this section the most important optimisations of the membrane installation are summarised.

Aeration in stand-by mode

In stand-by mode 50% of the capacity of one blower is used to aerate all tanks in a stand-by mode. The tanks are hereby aerated in rotation with an interval of 15 seconds. After, for instance, 10 of these cycles the aeration is stopped. Afterwards there is an interval when the aeration of the membrane streets is in a stand-by mode. After this interval the aeration cycle is started again.

The aeration control in stand by mode was set conservatively in the first six months. The pause time was set at time = 0 minutes, causing the cyclic aeration in the stand-by mode to be continuously on. Since the energy consumption of the membrane aeration constitutes a significant fraction of the total energy use of the installation, strategies to lower the membrane aeration were investigated. To achieve the above, two modifications were introduced in process operation. Firstly, on 1st June 2005 the aeration capacity of each blower was lowered from 4,500 Nm^3/h to 3,750 Nm^3/h. This reduction is justified since the aeration capacity was delivered for a system with full ZW500d cassettes each with 48 membrane elements. However in Varsseveld there were only 40 membrane elements placed per cassette. This modification led to a direct lowering of the aeration capacity by approximately 17%.

A second measure implemented was lowering the aeration capacity in stand-by mode. Within this measure from 16th June 2005 the pause interval, determining how long the aeration is off, was increased stepwise to 1 hour. The number of aeration cycles was set to 10. It means that each tank in stand-by mode is aerated for 15 seconds with a capacity of

1,875 Nm^3/h. This optimization led to a lowering of the energy consumption for membrane aeration by 30%.

Optimal flux

The optimal flux is the flux with which the membrane streets operate for a large part of time in process mode. The choice of the optimal flux determines therefore for a large fraction how many membrane tanks are in process- and in stand-by mode. At the increase of the optimal flux the switching between the membrane tanks would not take place so often, causing less membrane tanks to be in operation. The optimal flux determines therefore for an important part the energy consumption of the membrane filtration installation.

The optimal flux can be dictated by the plant operator. When selecting an optimal flux the optimal operation of the membranes is the most important. The operation of the membranes is evaluated based on the course of the permeability, both on long and short term. The short time permeability development, especially, is a good indication for the optimal flux. When the permeability within a permeate cycle decreases too much, or when the necessary/required pressure decreases too much, the performance is not optimal. Increase in the pressure means namely that the fouling of the membrane occurs, whatever the reason. The lowering of the optimal flux is then one of the possible measures. The optimal flux was changed few times at the WWTP Varsseveld as shown in Figure 4.3. From this figure it can be seen that the optimal flux is strongly related to the process temperature.

Figure 4.3 The course of the optimal flux and the process temperature

Membrane cleaning

In the first year of operation a lot of attention was paid to the optimization of the membrane cleaning. In the first instance the cleaning procedure was taken over from the simulation-unit; sodium hydroxide and citric acid were used. From April 2005 the sodium hydroxide was replaced by sodium hypochloride. In the spring and summer of 2005 the membranes were cleaned once to twice to three times a week, whereby good results were achieved. From autumn 2005 a little decrease in the permeability occurred and the cleaning frequency was increased, in accordance with the design, to once a week. In that period the cleaning procedure was further investigated.

In the optimisation route there were the following findings and the modifications introduced:

- The efficiency of a maintenance cleaning (MC) was worse when it was performed directly from the stand-by mode, especially in the winter. The reason for this was the presence of air in the permeate pipe, originating from the degasing of the saturated permeate. The MC was therefore performed only after degasing of the permeate pipe.
- During an MC in a completely empty tank mainly the lowest part of the membrane was cleaned. The MC's were therefore performed alternately at different water levels of the membrane tank.
- After the iron salt dosing in the aeration tank was increased, the inorganic fouling of the membranes increased. The dosing of the chemicals was then adjusted, whereby the duration of the acid dosing was increased. Because the duration of the acid- and chlor dosing were interrelated, the chlorine dosing was increased also. In the middle of 2006 the control software was adapted in a way that both chemicals could be set independently from each other.

4.4.3 Industrial polymer

For around half of February 2005 a strong reduction in the permeability occurred. In that period the average flux was high as a result of the influent of melt water and the process temperature decreased to below 10°C. Besides this, the influent of salt possibly played a role here and the fact that the aeration capacity was low as a result of problems with the aeration control. Both aspects had possibly a negative effect on the sludge quality. In one of the membrane streets, just in that period, the first peak test was performed. The course of the permeability in this membrane tank is given in Figure 4.4.

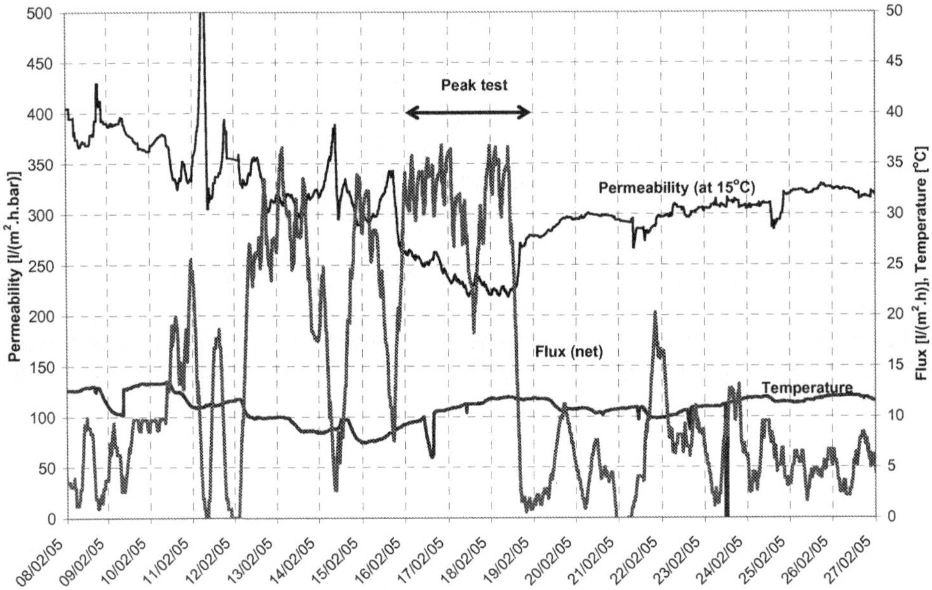

Figure 4.4.The permeability (corrected for 15°C) during a first peak test (16th -18th February 2005)

The permeability decreased within a short period from circa 375 to 225 l/(m^2.h). In the following 2 months the permeability stayed at the level of around 300 l/(m^2.h) and the trans-membrane pressure, especially at higher fluxes during the permeate discharge cycle, rose very fast. The reason for these problems was the sticking of bundles of membranes caused by deposition of a slimy substance (Figure 4.5). Due to that the membrane surface available for the permeate extraction was strongly reduced. With supplementary research carried out in the simulation unit it was found that this sticking was caused by an industrial polymer (see sub-research "Simulation-unit" [3]).

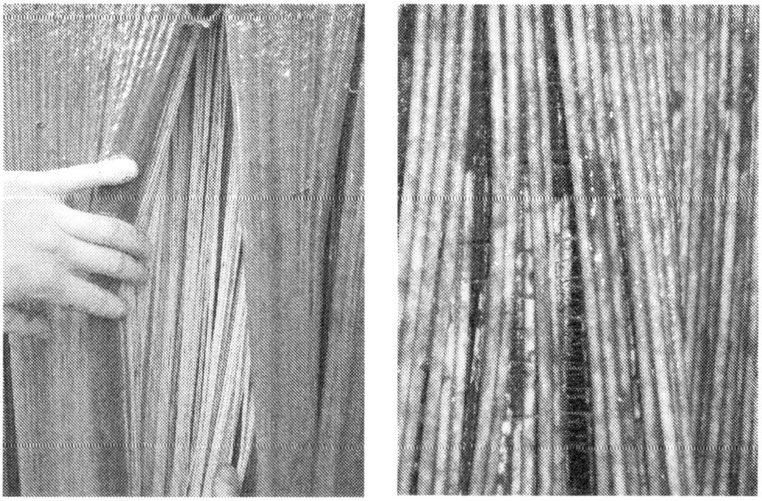

Figure 4.5 Sticking of the membranes of the simulation unit

The industrial polymer consists mainly of a co-polymer of vinyl acetate maleine acid di-n-butylester and is used, amongst others, as a cheese coating agent. This substance becomes hard when it comes in contact with air and is either very difficult to biodegrade or non-biodegradable. The local cheese factory discharges 5-10 m^3 wastewater containing this polymer in concentrated form. Research confirmed that the sticky layer contained the industrial polymer. Based on the van microscopic observation it was concluded that this sticking layer contained further sludge, filamentous organisms, fibres and large amounts of attached Ciliates. It is presumed that the industrial polymer stuck to the membranes and that this deposition offers a good attachment for the Ciliates. The slimy character of the deposition can be caused by the industrial polymer itself, the presence of the large colonies of Ciliates and by the microbial production of extra-cellular polymeric substances (EPS) (see sub-study "Sludge quality" [3]).

It was investigated in the laboratory and within the simulation unit as to how this sticking phenomena could be prevented. Based on attained information it was decided to intensively clean the membranes. This was performed in the middle of April 2005. Therefore sodium hypochloride was used in three membrane streets while for the fourth one hydrogen peroxide was used. The efficiency of chlorine was clearly higher than of hydrogen peroxide.

From the beginning of May 2005 the sub-stream containing the industrial polymer was disconnected and was transported to another WWTP. After these measures the permeability returned to the original level and the mentioned problems did not occur again.

4.4.4 Technical modifications

After the membranes were installed in November 2004 they were thoroughly inspected. Amongst others the membrane integrity was hereby investigated. Despite this, after a few months of operation the leakages occurred in different membrane cassettes. The leakages could be quickly detected using the turbidity meters in the four permeate pipes.

During the occurrence of leakage the sludge was discharged together with the permeate. This leads to a deterioration of the effluent quality. An even larger problem appears however, that at the regular back washing of the permeate, sludge is brought to the outer surface of the membrane. This can lead to fouling on the inner side of the membrane, and finally to a permanent deterioration of the capacity of the system.

The leakages turned out to be due to the cracks at the connection between some membrane modules and the permeate header. The membrane modules, where these problems occurred, were almost each time located on the outer side of a cassette. On the left photo of Figure 4.6 such a crack is shown. In the system where the modules were mounted in one cassette, the cracks also appeared in a few places (right photo in Figure 4.6). As a consequence of this a margin is generated in the fixing causing more moving freedom of the module. This could have contributed, amongst other things, to the cracks in the header connection.

Figure 4.6. Cracks in the header connection (left) and at the module fixing (right)

The (most probable) reason for the leakages was the alternating aeration mode (air-cycling) of the membrane cassettes. Because a relatively large amount of air (1,875 Nm^3/h) is switched between the cassettes in the membrane street within short time intervals (15 s), strong mechanical forces are imposed on the membrane cassettes. Especially the membranes on the outer side are subjected to these strong mechanical forces. These operational problems led to a change in aeration from the middle of 2005 when it was switched to sequential aeration of all membrane tanks instead of air-cycling. In sequential aeration the switch of air input does not take place between the cassettes but within the cassettes. The advantage of sequential aeration is the better distribution of the mechanical forces onto the cassette. From the technological point of view it decreases the efficiency of the membrane aeration and is not preferred.

Based on the experiences of the MBR Varsseveld, Zenon developed a technical modification in the construction of the cassette, which was mounted in March 2006 on the MBR Varsseveld. It comprised a better fixation hood, enabling the modules to keep in their place within a cassette. The old and new fixation hoods are shown in Figure 4.7.

Figure 4.7. The old (left) and the newly developed fixation hood (right)

To evaluate the efficiency of the new fixing hood the measurement of the forces within a module were performed in March 2006 on the MBR Varsseveld. The results were positive. By implementation of sequential aeration the new fixing hood keeps the module well in its place and significantly decreases the tensions within a module. When applying the cyclic aeration the forces within a module increase strongly. These forces are however much weaker than when the old fixing hoods were used. In the coming months the operation of air-cycling would be tested. After approximately 1 year the membrane modules will be examined again to evaluate the long term effect of the modification.

4.4.5 Macro-fouling

Despite all the measures in the design to prevent against macro-fouling of the membranes, in March 2006 macro-fouling was observed (Figure 4.8). In the upper 15 cm of the membranes the accumulation of hair and fibres occurred. Brush hair of the stationary micro-sieves (short of 0.8 mm thick) and from the scum layer scraper (longer of 1.2 mm thick) were found between the fibres. As reported in §4.2.4 the most probable reason for this fouling was incomplete sealing of the stationary micro-sieves.

Figure 4.8 Macro fouling of membranes (left: fibres in the upper part, right: sludge pieces in the lower part)

Under the membranes agglomerates of anaerobic sludge were found. It is known that the aeration does not reach the lowest part of the membranes; this phenomenon was also observed at the other locations with Zenon membranes. The fact that it is hardly aerated during the stand-by mode could also contribute to the anaerobic conditions. The membrane tank is always oxygen deficient for a large time during the stand-by mode.

The oxygen concentration in the membrane tank during stand by mode is shown in Figure 4.9. The aeration frequency during the stand-by mode was dependant on the number of tanks being in the stand-by mode. The aeration is changed between the tanks in stand-by mode. In Figure 4.9 there is a distinction made between situations when 1, 2 or 3 membrane tanks are in stand-by mode. Also the oxygen course in a membrane tank in operation mode is given. The plots are based on the model calculations whereby the process settings and the average process conditions of 2005 served as starting points.

To prevent against formation of anaerobic sludge fragments the aeration regime had to be adjusted. The following options were hereby considered:

- Applying cyclic- instead of sequential aeration, causing periodical doubling of the maximal aeration capacity within a membrane cassette and increase of turbulence around the membranes;
- Regular (e.g. once a week) long lasting aeration during stand-by mode to prevent against formation of sludge fragments and/or to remove the present sludge flocs;
- Adjustment of process control in a way where directly after the stand-by mode no permeate discharge directly takes place (in oxygen deficient sludge) but is aerated first for a short time.

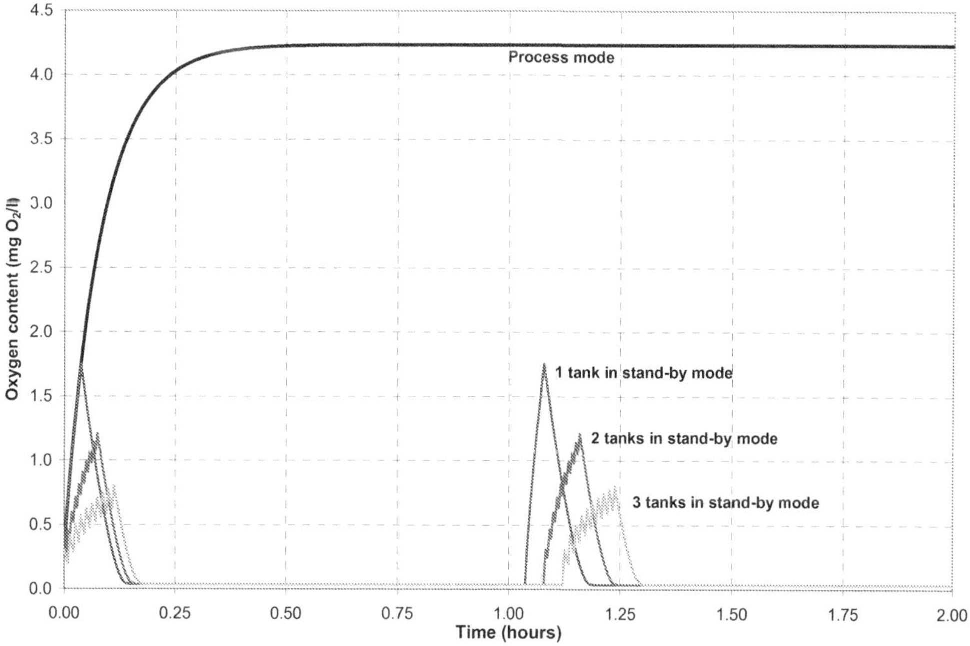

Figure 4.9 The oxygen concentration in the membrane tank in operation- and stand-by mode (based on model calculations)

5

Interactions

5.1 INTRODUCTION

In the previous chapters the design, the results and the experiences with the operation of the MBR Varsseveld were discussed. In this chapter the focus is on all the relations between these aspects. Based on them new insights were obtained and improvements for design and operation could be identified.

In §5.2 the aspects having a link with sludge quality are described. The excess sludge production is discussed in §5.3. In §5.4 the phosphate balance over the WWTP is made and interpreted in which rate the biological and chemical processes played a role. The nitrogen balance is described in §5.5 and the oxygen balance in §5.6. Hereby the influence of aeration on the membrane tank is discussed. In §5.7 the energy consumption of the whole installation is described per process unit. The aspects related to the membranes are finally discussed in §5.8.

5.2 SLUDGE QUALITY

5.2.1 Introduction

In the last years the opinion was that a good performance of membrane filtration is for the most part determined by the sludge quality. The sub-study "Sludge quality" [3] is completely focussed on the relation between the properties of the sludge in the MBR Varsseveld and the filterability of the sludge. The goal of this was to get a better understanding of the complex interactions between the sludge quality and the filterability and based on this to formulate optimisations in the biological process operation to reduce the membrane fouling. In the framework of this sub-research an investigation on several levels was performed into the sludge properties and is summarised in §5.2.2. The occurrence of the scum layer is further analysed in §5.2.3. Finally in §5.2.4 the relation between the sludge properties and the filterability of the sludge are discussed.

5.2.2 Microscopic examination

The microbial composition of the activated sludge population is determined by a number of factors, such as sludge loading and sludge retention time, the composition of the wastewater, the oxygen concentration and the aeration regime, the configuration of the installation and the process conditions (temperature and pH). The start-up of a new treatment installation with sludge from another installation can lead to a change of one or more of these factors, resulting in a shift in the activated sludge population. Regarding large differences between the process performance of the MBR Varsseveld and of the 'old' high loaded conventional installation the consequences of the switch on the microscopic picture were remarkably small.

The most remarkable was the change of the floc structure: The sludge flocs became over the course of time smaller and less compact (Figure 5.1). This could result from:

- High turbulence. Turbulence can lead to the damage of the flocs and their (partial) disintegration into smaller flocs;
- The absence of a sedimentation step. Bacteria which do not flocculate would hardly remain in a conventional installation and would be washed. In an MBR the necessity of flocculation is smaller, because the membranes prevent against wash out. This would theoretically lead to a structurally larger fraction of free cells in the sludge, which did not occur.

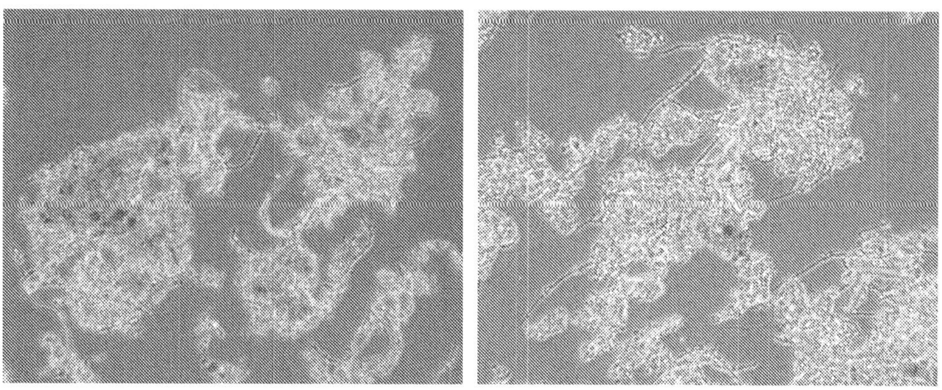

Figure 5.1. Microscopic picture of sludge (100x) during the start-up (left, 4th January 2005) and after 2 months of operation (right, 28th February 2005)

A second remarkable change was the increase in the number of bio-P colonies. This aspect is further discussed in §5.4.2. The other observations during the microscopic sludge examination were not structural enough that they could be attributed specifically to the MBR-configuration.

5.2.3 Floating (scum) layer

The formed scum layers as described in §4.3.4 were examined under microscope (sub-study "Sludge quality" [3]). When the scum layer appeared for the first time (February 2005) a larger population of *Microtrix parvicella* was observed in the sludge. This is a filamentous micro-organism stimulating flotation which in communal wastewater treatment plants with nutrient removal is considered as the main cause of scum layer forming and bulking sludge. It is reasonable to suppose that this scum layer was (partially) caused by these bacteria. It is notable that a scum layer was formed when the filamentous population was not too large. The same counts for the scum layers in July and October 2005: also then there was not extremely much filamentous present in the sludge and the filament index was relatively low (1-2). In communal installations the chance for the scum layer to occur is bigger when a filament-index is 3 or higher, whereby moreover, one or more of the flotation causing filamentous organisms are dominant. In July there was a population of Actinomycetes observed but not in such quantities that the scum layer would be expected. In October there were both *M.parvicella* and Actinomycetes present.

The above considerations allow one to suppose that other factors played a role in scum layer formation. It is hereby mainly thought that a (periodical) increase of the concentration of hydrophobic compounds in the influent was responsible for that phenomenon. This assumption is based on the fact that the wastewater sometimes contained a strongly increased concentration of compounds which caused the sludge to become (periodically) more hydrophobic. On the day before the extreme scum layer formed in July 2005 there was a very high COD content measured in the influent (>2,000 mg/l). It is considered possible that some compounds which were retained by the membrane and accumulated in the sludge caused more hydrophobic character than the 'average' activated sludge from a conventional installation. In that case a negligible increase in the number of filamentous organisms would be sufficient to cause the sludge floating.

This hypothesis seems to be confirmed by the measurements of the hydrophobicity of the sludge. In Figure 5.2 the hydrophobicity is plotted against the sludge volume index (SVI) and the observed scum layers. It is remarkable that the problems with scum layers became serious just after the measured hydrophobicity increased. Also during the scum layer appearance in July and October 2005 a clear, sudden increase of the SVI was measured.

To test the influence of the MBR-sludge and the MBR influent on the flotation affinity of the sludge some flotation tests were performed. Hereby it turned out that the sludge concentration had an influence mainly on the flotation phenomena. The higher the sludge content was, the higher the sludge fraction that flocculated. The influence of the influent composition could not be proved.

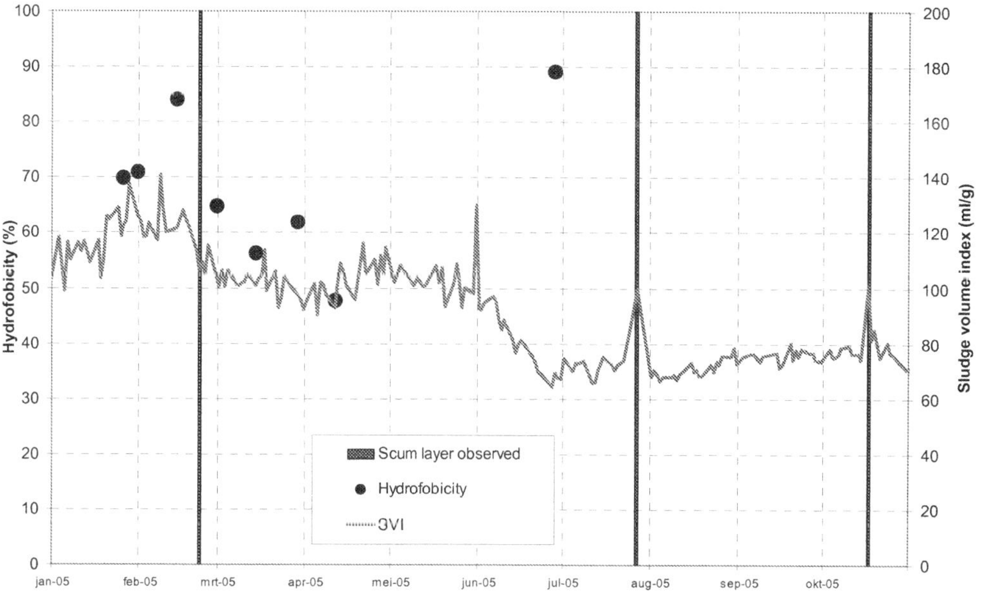

Figure 5.2. The forming of flotation layer in relation to the hydrophobicity and the SVI of the activated sludge

The exact reason for the flotation layer forming at the MBR Varsseveld did not become clear during this research. It is assumed that the following factors could have played an important role:

- the high dry matter concentration of the sludge, causing the sludge to be more susceptible for the flotation,
- the relatively large amount of air brought in the aeration circuit and the membrane tanks;
- a structural higher hydrophobicity of the sludge as a consequence of the accumulation of the hydrophobic compounds,
- a temporary provision of hydrophobic components with the influent.

5.2.4 Filterability

Membrane fouling in MBR-systems is a complex phenomena determined by a combination of factors. Hereby these can be thought about as the composition of the sludge mixture to be filtered, the membrane characteristics (amongst others the size of the pores and the material of the membrane), the construction of the membrane modules and the way in which these modules were operated. A large number of sludge properties can play a role by membrane fouling. The sludge concentration, the sludge composition, the size of the flocs, the strength of the sludge flocs, the presence of multivalent cations and the concentrations and composition of the dissolved and colloidal matter in the water are all of importance. Since these factors, moreover, are related to each other, it makes establishing the relationship between sludge quality and the membrane fouling even more complex.

During the period January until October 2005 the filterability of sludge samples and fractions of these samples from the MBR Varsseveld were determined. Figure 5.3 shows the filtration curves as measured using a test set-up at a flux van 60 l/(m^2.h). In this figure the membrane resistance is given against the amount of filtered medium. A quick increase of the resistance means that the membrane becomes clogged quickly.

A clear distinction between the sludge samples taken in the period January-April (with the exception of 14th February) and the samples taken in June and October can be made. With the sludge samples from the period January - April 2005 the resistance increased quickly and the filterability of the sludge samples were evaluated as 'bad'. In the sludge samples taken in June and October the resistance increased much more slowly and the filterability of the sludge samples was evaluated as "good". The cutting off of the industrial polymer at the beginning of May had a large influence on the improvement of filterability of the sludge.

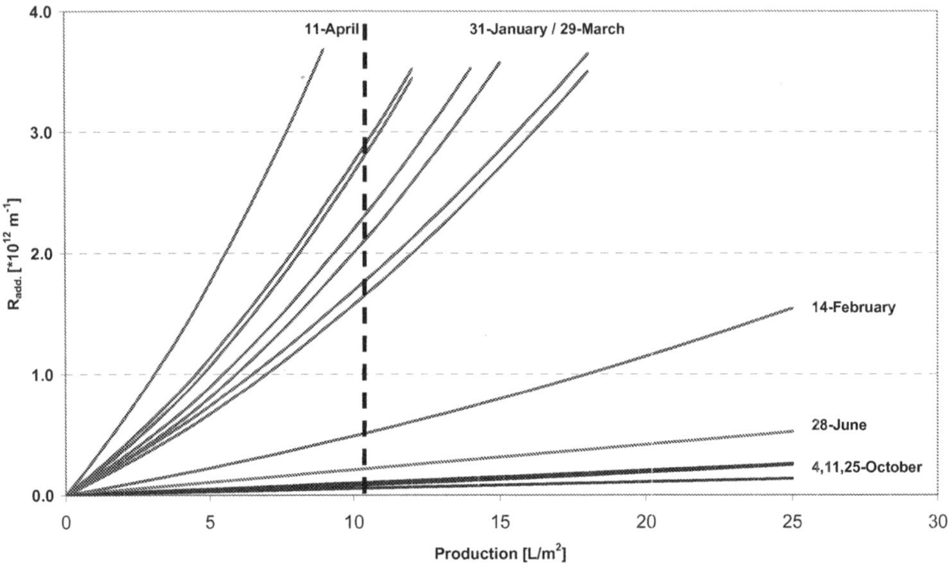

Figure 5.3. Filtration curves of sludge samples from the membrane tank of the MBR Varsseveld (2005)

Also there were different physical and chemical parameters of the sludge samples measured and the relationships were established between the filterability of the sludge and the results of these analyses. From that it turned out that the particles in the "free" water or "supernatant", and more specifically the particles in that supernatant being larger than 0.45 µm (the "colloidal" fraction), played the most important role in the filtration resistance of the sludge. From a more detailed characterization of the supernatant it seems that especially high concentrations of "colloidal" polysaccharides (part of EPS) were associated with a poorer filterability of the sludge. These particles are most probably not brought with influent wastewater but produced during the biological treatment process.

From above considerations it can be concluded, from the point of view of the biological process, that in order to reduce the membrane fouling in MBR systems, the attention has to be focused on maintaining the content of colloidal material in the sludge mixture as low as possible. Hereby the relationship between flocculation and deflocculation of the sludge flocs plays an important role. The factors having influence on these comprise the mixing intensity and oxygen concentration. A lower mixing intensity and a higher oxygen content have in general a positive effect on the floc forming and lower the concentration of colloidal particles.

From the strongly positive correlation found between the hydrophobicity in the sludge flocs and the filterability of the sludge proved the importance of the binding strength. A higher hydrophobicity result in a stronger binding makes the flocs less sensitive to de-fragmentation and de-flocculation.

5.3 SLUDGE PRODUCTION

The excess sludge production on the MBR Varsseveld can be quantified in different manners. In this section the results of various methods are compared to each other. The sludge production is subsequently calculated based on:

- The data on the sludge transport (§3.3.5). Hereby an estimation is made for the contribution of the sieve goods and the grease. The resultant is the sum of the excess sludge and the scum layer discharge (the latter can also be considered as excess sludge);
- The operation hours of the thickened sludge pumps and the dry matter measurements in the thickened sludge. A certain correlation was assumed hereby between the electricity consumption of the pumps and their capacity. The excess sludge discharged with the scum layer via the grease drain was not accounted for in this method;
- Based on the phosphate balance (§5.4). This represents the excess sludge including the scum layer;
- Based on the theoretical static model calculations (method of Chudoba). This represents the excess sludge including the scum layer.

The results of the four methods are given in Table 5.1.

The correspondence between the different methods was relatively good, taking into account the accuracy. The phosphate balance leads to the highest values. In the phosphate balance calculations phosphate removal efficiency in the pre-treatment was taken as high as 8%, as measured during the research (§3.2.2). The variations in these measurements were quite significant (from –8 to 22%). When in the phosphate balance the reduction efficiency of 20% was assumed the total sludge production decreases from 1,380 to 1,170 kg MLSS/d.

Table 5.1 Sludge production on the WWTP Varsseveld estimated with four different methods

	Method excess sludge determination	Unit	Conventional	Total	Phase 2	Phase 3	Phase 4	Phase 5
	Ferric dosing	*mol Fe/mol $P_{influent}$*	*0.5*	*0.53*	*0*	*0.40*	*0.82*	*1.12*
1	Discharge data	kg MLSS/d	1,320	1,030	1,090	810	1,170	890
2	Thickened sludge pumps[1)]	kg MLSS/d	-	1,060	1,010	910	1,070	1,270
3	Phosphate balance	kg MLSS/d	-	1,380	1,340	1,530	1,490	1,360
4	Model calculation	kg MLSS/d	1,360	1,050	900	1,040	1,120	1,080
	Average	kg MLSS/d	1,340	1,130	1,090	1,070	1,210	1,150

1) Based on the estimated dry matter content in the scum layer drain of 100 kg MLSS/d

The theoretically calculated amount of sludge corresponds well with discharge (transport) data for both, the MBR and the old WWTP. Based on the results of both methods it can be concluded that the excess sludge production of MBR Varsseveld is significantly lower than of the old conventional installation. The difference in the sludge production between both systems is a consequence of the following factors:

- The loading of the conventional installation in 2004 was higher than that of the MBR in 2005;
- The sludge loading of the conventional installation was significantly higher than that of the MBR;
- The pre-treatment of the MBR removes a part of the pollution load, in the conventional installation there was no pre-treatment present;
- The average iron salt dosing of the MBR was somewhat higher than that of the conventional installation;
- At the conventional installation a part of the sludge was washed out with the effluent.

The influence of these five aspects was investigated using model calculations. The results are given in Table 5.2 (the shaded columns are based on the full-scale data from 2004 and 2005, the remaining columns are derived from them). As a starting point for the calculations the average pollution load from the total research period was taken. Hereby distinction was made between MBR and conventional WWTP. For the MBR the sludge production of the whole research period is considered and the situation whereby the iron dosage is so high that the permeate target value for phosphate (0.15 mg P_{total}/l) can be reached. For the conventional WWTP the situation from 2004 is considered and the following two fictive situations: the high loaded WWTP at the loading from 2005, and the transformation to a conventional system with effluent filtration, with which the target effluent values can be reached.

Table 5.2. Sludge production (in kg MLSS/day) of the (old) conventional WWTP and the MBR Varsseveld

Type installation Unit		Conventional WWTP Varsseveld			MBR Varsseveld	
Year		2004	2005	2005	2005	2005
Loading	p.e. (a 136 g TOD)	29,700	31,500	31,500	31,500	31,500
Effluent concentration	mg P_{total}/l	1.6	1.6	0.15	2.2	0.15
Ferric dosing	l/h	12.1	12.1	22	11.6	22
Sieve goods	(1)	0	0	0	200	200
Biological excess sludge	(2)	1,260	1,470	1,190	910	920
Chemical excess sludge	(3)	130	140	220	140	240
Wash out	(4)	-30[1)]	-30[1)]	-10[2)]	0	0
Excess sludge produced	(2+3)	1,390	1,610	1,410	1,050	1,160
Excess sludge discharged	(2+3+4)	1,360	1,580	1,400	1,050	1,160
TOTAL discharge	(1+2+3+4)	1,360	1,580	1,400	1,250	1,360

1) With the influent flow of 4,480 m^3/day and a suspended solids of 8 mg/l in 2004

2) With the effluent filtration using sand filters and a suspended solids concentration of 2.5 mg/l

Also based on these considerations it can be concluded that the excess sludge production of the MBR (1,160 kg MLSS/d) is lower than that of the conventional WWTP at the same sludge loading rate (1,410 kg MLSS/d). This is a consequence of the fact that in the MBR there was a pre-treatment including micro sieves, causing higher removal of COD, total solids (TS) and other parameters than in conventional WWTP. When the sieve goods are incorporated in the calculation of the excess sludge production, the total excess sludge discharge of both systems is similar.

5.4 PHOSPHATE BALANCE

5.4.1 Introduction

The influent wastewater of the WWTP Varsseveld has a relatively high phosphate concentration of 15 mg/l on average base. The effluent requirement of the WWTP is in the summer equal to the MAR (maximally acceptable risk) norm van 0.15 mg P_{total}/l. Reaching this effluent quality is not simple. In the MBR Varsseveld the phosphate removal takes place through a combination of biological and chemical processes. There was no anaerobic tank designed, but because of the low nitrate concentration in the pre-denitrification tank, the conditions were created there, where the development of the phosphate accumulating organisms (PAOs) took place. The largest fraction of the phosphate needs to be removed, however, via chemical precipitation, with dosing of ferric choride sulphate ($FeClSO_4$). In this section the results of the phosphate removal at the MBR Varsseveld are described.

5.4.2 Biological phosphate removal

In the first months after the start-up there was no iron salt dosed. The goal of this was to give the biological phosphate removing bacteria the chance to develop and to get an insight into the possibilities of biological phosphate removal. From the end of May 2005 the iron salt dosing was started and its amount was slowly increased. The goal of this was to prevent against the washout of PAOs.

The presence of biological phosphorus removal was found using the following methods:

- Phosphate release tests on the 25th May 2005 and 28th April 2006;
- Microscopical analysis (sub-study "Sludge quality" [3]);
- Modelling of the MBR Varsseveld (sub-study "SIMBA modelling" [3]).

The results of the different methods are summarised in Figure 5.4.

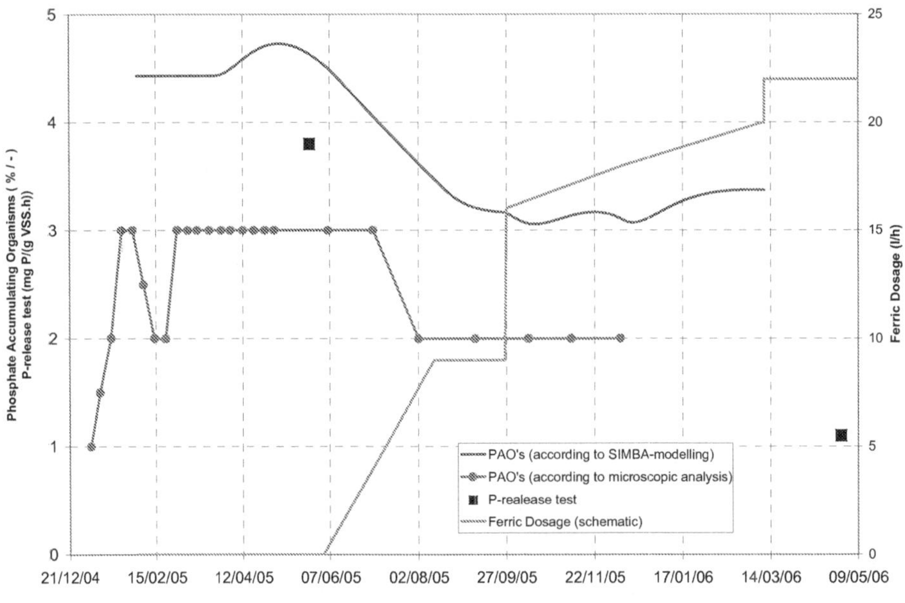

Figure 5.4. The course of the biological phosphorus removal

Based on the microscopic examination no quantitative comparison can be in principal made but a trend can be sketched. At the start-up there was a limited amount of PAOs colonies present (score 1); within a number of weeks this amount increased to a score of 3. This population remained present for a few months but it decreased after the ferric dosing started to a score of 2. Based on the three applied methods the same conclusion can be drawn in principle: as a consequence of the ferric dosing the biological phosphate removal efficiency decreases.

The strong decrease in the phosphate release is remarkable. In May 2005 it was still satisfactory according to the STOWA methodology [6]. A year later, after the ferric dosing increased significantly and the phosphate concentration decreased to 0.15-0.20 mg P_{total}/l, some biological phosphate removal still occurred.

5.4.3 Chemical phosphate removal

For the different research periods a phosphate balance was set. Based on phosphate balance the relationship between chemical and biological phosphate removal was determined. The results of this are given in Table 5.4. As a starting point it was assumed that the efficiency of the ferric dosing decreases when the Metal/P ratio increases and the phosphate concentration in the permeate decreases.

Based on the phosphate balance it can be stated that the biological phosphate accumulation decreased from approximately 2.7% to 1.9% dry matter. The chemical bound phosphate fraction increased to 2.0%. When the biological phosphate accumulation based on the biological sludge is calculated, this fraction decreased from 2.7 to 2.2-2.3%. Also based on the phosphate balance it can be concluded therefore that the biological phosphate accumulation decreased as a result of the ferric dosing. The amount of biologically accumulated phosphate is however so high that the biological phosphate accumulation still occurred. This corresponds with the results as presented in §5.4.2.

In the winter of 2005/2006 the ferric dosing was adjusted to reach the target values of 0.15 mg P_{total}/l. This target value counts however only for the months April to October. In the future therefore there will be a higher dosing necessary during these 7 months (Me/P-ratio of circa 1.2). In the winter months a lower dosing (Me/P-ratio of circa 0.6) is applied. In the summer months the biological phosphate removal capacity will decrease, in the winter months it will again increase. Summarising it can be stated that the biological dephosphating, also at strength effluent requirements as of Varsseveld, play an important role.

Table 5.4 The phosphate balance of the different research periods

Parameter		Unit	Total	Phase 2	Phas e 3	Phase 4	Phase 5
Phosphate balance	Influent	kg/d	65.2	60.8	70.1	67.1	59.0
	Permeate	kg/d	-8.6	-19.1	-9.7	-2.0	-0.8
	Pretreatment[1)]	kg/d	-5.2	-4.9	-5.6	-5.4	-4.7
	Removal	kg/d	51.3	36.8	54.7	59.7	53.5
Ferric dosing	Dosing flow	l/h	11.6	0.0	7.4	17.4	21.8
	Iron loading	kg Fe/d	52	0	34	79	98
	Me/P ratio	mol Fe/mol P_{inf}	0.53	0	0.40	0.82	1.12
	$FePO_4$ fraction[2)]	%	75	-	100	70	50
Sludge production	Total	kg MLSS/d	1,380	1,340	1,550	1,490	1,360
	Chemical	kg MLSS/d	130	0	90	190	230
	Biological	kg MLSS/d	1,250	1,340	1,460	1,300	1,130
P-accumulation	Biological	kg P/d	29.5	36.6	36.1	29.2	26.2
	- % total sludge	% P	2.1	2.7	2.3	2.0	1.9
	- % biological sludge	% P	2.4	2.7	2.5	2.2	2.3
	Chemical	kg P/d	21.8	0.0	18.6	30.5	27.3
	- % total sludge	% P	1.6	0.0	1.2	2.0	2.0
	Total	% P	3.7	2.7	3.5	4.0	3.,9

1) Assuming a phosphate removal of 8% (§3.2.2)

2) The fraction Fe that is bound to $FePO_4$ (and not $Fe(OH)_3$) is dependant of the phosphate content in the aeration circuit

5.4.4 Organic phosphate

During the degradation processes in the activated sludge, soluble organic macro-molecules are released/produced such as humic acids. These are found back in the permeate as COD. Organic fractions of nitrogen and phosphate are bound to the humic acids. The organic phosphate content in the permeate of the MBR Varsseveld varied in practice between 0.02 and 0.04 mg P/l when ferric dosing was applied. In the situation without ferric dosing the organic phosphate concentration in the permeate was however significantly higher and varied between 0.4 and 0.6 mg P/l. A part of the organic bound phosphate is removed chemically after ferric dosing. The permeate value for phosphate amounted 0.15 mg P/l. Based on the above considerations it can be concluded that this target value could not be reached on the MBR Varsseveld without chemical dosing. At the implementation of ferric dosing the organic bound fraction constitutes still around 10-20% of the target value and reaching the target value is possible.

In the research at the MBR Hilversum [7] it was attempted to reach the permeate concentration of 0.15 mg P_{total}/l only via the biological way, thus without addition of iron salts. The organic phosphate concentration of the permeate of Hilversum amounted to 0.14 mg P_{total}/l on the average base. This means that a permeate concentration of 0.01 mg P_{ortho}/l needs to be reached to attain this requirement.

The influence of the organic phosphate concentration on the feasibility of the effluent requirement is different per location, and depends mainly on the influent composition and the sludge loading. At a too low sludge loading more humic acids are formed and therefore more organic phosphate is bound. A relatively low rate dosing of iron salts can lead to a significant decrease of the organic phosphate content and therefore can simplify the reaching of the effluent requirements.

5.5 NITROGEN BALANCE

The results have shown that the nitrogen removal at the MBR Varsseveld complied easily with obliging the required standards. The target value of 5.0 mg N_{total}/l was reached for the largest part of the year. The nitrogen balance of the different research periods is given in Table 5.5.

Table 5.5 The nitrogen balance of the different research periods

Parameter		Unit	Total	Phase 2	Phase 3	Phase 4	Phase 5
N_{Kj}	Influent after pre-treatment	kg/d	192	193	199	189	180
	Permeate	kg/d	-4.2	-4.1	-3.6	-4.2	-5.3
	Discharge with sludge	kg/d	-77	-78	-91	-81	-78
	Nitrification capacity	kg/d	112	111	104	103	97
	Efficiency nitrification	kg/d	98	98	98	98	97
NO_3-N	Influent after pre-treatment	kg/d	8.1	16.5	4.0	7.3	7.9
	Permeate	kg/d	-11.2	-8.5	-9.9	-15.5	-10.8
	Denitrification capacity	kg/d	108	119	98	95	94
	Efficiency denitrification	%	91	93	91	86	90

The nitrification efficiency of the biology was 98%. This is relatively high for a biological treatment. This high reduction became possible because of the intensive aeration in the membrane tank. In this last process step, the ammonia present is oxidised to a large extent. The consequence was very low permeate concentrations of 0.2-0.4 mg NH_4-N/l.

The high denitrification efficiency was achieved thanks to the beneficial COD:N ratio in the influent. In the design a COD:N ratio of 10.7 was assumed. In practice it varied however between 12 and 13. The COD:N ratio of the influent after pre-treatment was 10-11.

The organic nitrogen content in the permeate varied in practice between 0.6 and 1.2 mg N/l. For the MBR Varsseveld it means that the fraction of non-biodegradable organic nitrogen constitutes approximately 10-25% of the allowable nitrogen discharge of 5.0 mg N_{total}/l. When the effluent requirements became even more stringent, for instance 2.2 mg N_{total}/l, this fraction would increase to 25-50% and would play an important role. The space that remains then for other nitrogen components becomes very small. The ferric dosing seems not to have any influence on the organic nitrogen content in the permeate.

5.6 OXYGEN BALANCE

5.6.1 Introduction

The oxygen balance of an MBR is different from that of a conventional WWTP. Because of the high dry matter content the oxygen diffusion is worse, thus more oxygen has to be provided into a smaller reactor volume. In addition oxygen is brought also into the membrane tanks, although its efficiency here is not too high. In the sub-study "OC and hydraulics" [3] there were measurements performed on the oxygen balance in both the aeration circuit and the membrane tanks. In this section the most important results are presented and the interactions between the aeration circuit and the membrane tanks are analysed.

The oxygen input to the aeration tanks and the membrane tanks are described in §5.6.2 and §5.6.3. In §3.3.6 based on, amongst other things, measurements of the α-factor, an oxygen balance is established.

5.6.2 Oxygen input aeration tank

In §2.4.3 it was emphasized that the design of the aeration deserved special attention. In the framework of the sub-study "OC and hydraulics" [3] therefore, the oxygen input capacity of the aeration in the circuit is regularly measured. The goal of this was to define the initial situation and to monitor how the aeration capacity developed in time. The results of these measurements are summarised in Table 5.6.

Table 5.6 Results of the OC-measurements in clean water and activated sludge

Date	Unit	18-Nov-04	24-Jan-05	14-Jun-05	12-Dec-05	22-Dec-05
Medium	-	Water	Sludge	Sludge	Sludge	Sludge
Installed capacity of the impellers	kW	3.0	3.0	4.0	6.5	4.0
Temperature	°C	10.3	10.5	18.5	13.0	13.0
α-factor	-	1.02	0.76	0.58	0.71	0.71
Flow velocity	cm/s	16.7	16.7	22.7	29.0	22.7
OC(10)	kg O_2/h	191	183	114	156	124
OC specific	g O_2/(Nm^3.m)	18.6	18.6	12.1	15.5	12.2

In November 2004 there were guarantee measurements performed in the aeration tank filled with ground water. Here it was remarkable that the measured OC-value of 191 kg O_2/h was significantly lower (30%) than the design value of 273 kg O_2/h. The specific oxygen input amounted to 18.6 g O_2/(Nm^3.m) instead of a guaranteed specific input of 27 g O_2/(Nm^3.m). This result was confirmed during the first OC-measurements in sludge, the 24th January 2005. A possible reason for the low oxygen input could be a low flow velocity in the aeration circuit. The flow velocity amounted to 16.7 cm/s, while according to the specified requirements 30 cm/s should be reached.

Another reason was the optimistic design guideline used by the manufacturer of the aeration system. In Figure 5.5 the design parameters and the results of the guarantee measurements are plotted against the results of OC-measurements from other locations using the same type of aeration plates. From this it can be seen that the specific aeration input as used in the design was significantly higher than the results attained in other WWTPs. It is remarkable that the measured specific oxygen input in Varsseveld was lower than measured at the other WWTPs. This was because of the low flow velocity and less favourable configuration.

To improve the oxygen input the installed capacity of the impellers was increased at the end of January 2005 from 3.0 to 4.0 kW. On the 12th December 2005 there was even a test performed with a temporary setting to 6.5 kW. The results of the OC-measurements in June and December 2005 showed however a significantly lower oxygen input than in January 2005. The oxygen input decreased by 35%. In comparison with the design the practical oxygen input was 55% lower. Implementation of the impellers of 6,5 kW seemed to have a positive effect on the oxygen input; the capacity increased by 25-30%.

The result of the strong decrease of the oxygen input capacity of the plate aerators is currently further investigated. In April 2006 three aeration plates in the aeration circuit were taken out of operation and further research was performed on site by the manufacturer. Hereby the bubble pattern was, amongst others, judged visually and the pressure measurements were carried out. Based on these tests it was concluded that the plates were not fouled. Despite the fact that there were more indications that the OC decreased dramatically, the reasons for this are still not clear.

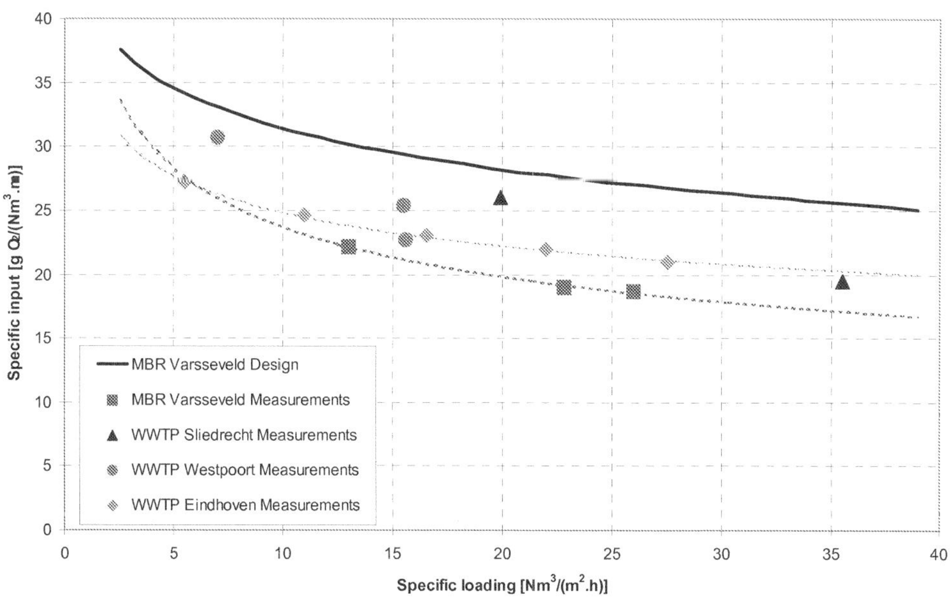

Figure 5.5. The specific oxygen input in relation to the plate loading for different WWTPs

5.6.3 Oxygen input membrane tanks

In the framework of the sub-study "OC and hydraulics" [3] the measurements were performed on the oxygen input in the membrane tanks. Depending on the influent flow to the membrane tanks the oxygen input α-OC(10) varied between 10.4 and 18.0 kg O_2/h, resulting in a specific oxygen input of 2.0 to 3.4 g $O_2/(Nm^3.m)$.

Based on these measurements a sensitivity analysis was performed where the influence of the oxygen concentration in the aeration tank (influent to the membrane tank) on the oxygen content in the membrane tank was determined. To attain it a theoretical model was set where all effects (influent, discharge, respiration and aeration) were included. The results of the sensitivity analysis are graphically presented in Figure 5.6. Hereby a constant respiration rate of 2.5 mg O_2/(g MLSS.h) was assumed and the process parameters were the same as those set at the MBR Varsseveld from the middle of 2005 (aeration membranes of 3,750 Nm^3/h, influent membrane tanks of 800 m^3/h and an optimal flux of 20 $l/(m^2.h)$).

Based on Figure 5.6 it can be concluded that the influence of the oxygen concentration in the influent to the membrane tanks had a limited effect on the oxygen content in the aeration tank. An increase of the oxygen concentration in the influent of 1.0 mg O_2/l leads to an increase of the oxygen concentration in the membrane tank of 0.3-0.4 mg O_2/l. The aeration capacity of the membrane is sufficient, when the degradation in the aeration tank proceeds well and the respiration level of the sludge is low, the oxygen content in the membrane tank is then always high enough. Based on the world-wide experience with MBR-installations it is always stated that it is undesired when membranes extract permeate when the sludge is under anaerobic conditions. This can lead to an accelerated fouling of the membranes and deterioration of the permeability. From the above sensitivity analysis however it can be concluded that a good biological conversion in the aeration tank is more essential than feeding with oxygen rich sludge mixture.

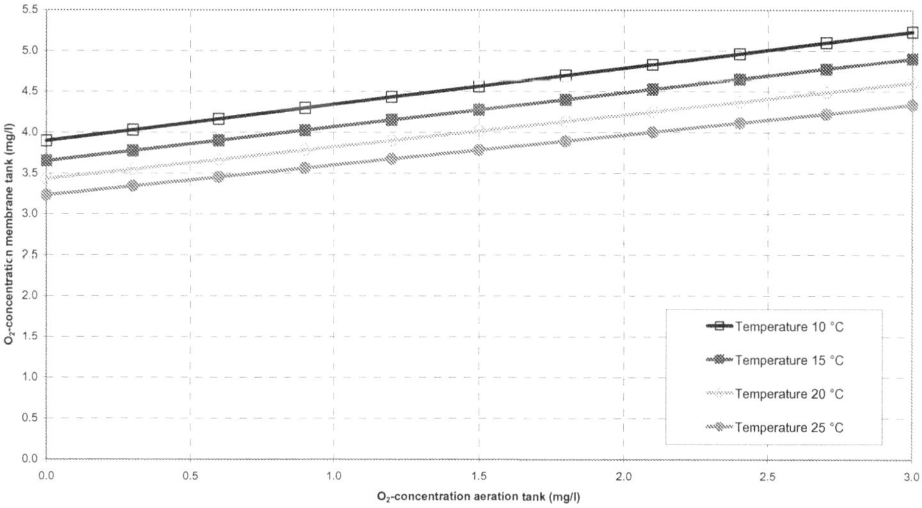

Figure 5.6. The influence of the oxygen concentration in the aeration tank on the oxygen content in the membrane tank

5.6.4 Oxygen balance

Based on the attained data for various research periods an oxygen balance was set-up, as given in Table 5.7. Hereby the specific oxygen input for the aeration in the circuit and the membrane tank as measured in the sub-study "OC and hydraulics" [3] were used. For the aeration in the membrane tank the α-OC value was used. It was assumed that for the coarse bubble aeration in the membrane tanks the α-factor was equal to 1 [8]. Finally, the theoretical α-factor was calculated based on the oxygen balance and compared with the measured α-factor.

The calculated α-factor for the aeration in the circuit turned out to be circa 25% lower than the measured values. This is a significant difference that cannot be explained based on the available data. In the balance a number of factors were taken into account which have a significant influence on the result. This comprises, amongst others, the oxygen input capacity in the circuit, the efficiency of the pre-treatment and the α-factor measurement.

The calculation of the oxygen input in the aeration circuit is based on an efficiency of 12.1 g $O_2/(Nm^3.m)$, as measured in June and December 2005 (see Table 5.6). When an efficiency of 18.6 g $O_2/(Nm^3.m)$ is presumed, as determined during the guarantee measurements in November 2004, then the calculated α-factor for the total research period would decline from 0.50 to 0.33 and the difference in the calculated and measured values would increase from -25% to –51%. The results of the oxygen balance confirm the conclusion that the oxygen input capacity in the aeration circuit has decreased significantly since the guarantee measurements.

Table 5.7 The oxygen balance over the different research periods

Parameter		Unit	Total	Phase 2	Phase 3	Phase 4	Phase 5
COD	Influent aeration tank	kg/d	2,368	2,328	2,501	2,376	2,162
	Discharge with permeate	kg/d	-92	-99	-90	-90	-88
	Discharge with excess sludge	kg/d	-1,301	-1,380	-1,406	-1,308	-1,233
	Oxygen demand *(a)*	kg/d	975	849	1.005	978	840
Nitrogen	Nitrification capacity	kg/d	112	111	104	103	97
	Denitrification capacity	kg/d	108	119	98	95	94
	Oxygen demand *(b)*	kg/d	200	166	195	200	174
Aeration circuit	Aeration capacity	Nm^3/h	1,250	1,319	1,594	1,092	949
	Efficiency	$g\ O_2/(Nm^3.m)$	12.1	12.1	12.1	12.1	12.1
	Depth aeration	m	5.1	5.1	5.1	5.1	5.1
	Oxygen concentration	$mg\ O_2/l$	1.5	1.2	1.3	1.7	1.7
	Temperature	°C	15.6	14.6	20.3	13.9	11.5
	Oxygen saturation	$mg\ O_2/l$	9.9	10.1	9.0	10.3	10.9
	Oxygen input *(c)*	$kg\ O_2/d$	1,574	1,716	2,022	1,349	1,183
Aeration membrane tanks	Aeration capacity	Nm^3/h	4,409	5,849	3,768	4,086	4,100
	Efficiency (α-OC(T))	$g\ O_2/(Nm^3.m)$	2.1	2.1	2.3	2.1	2.0
	Depth aeration	m	2.8	2.8	2.8	2.8	2.8
	Oxygen concentration	$mg\ O_2/l$	3.8	3.6	3.4	3.6	3.8
	Oxygen input *(d)*	$kg\ O_2/d$	383	532	362	375	359
OC	Oxygen demand total $(e = a + b)$	$kg\ O_2/d$	1,175	1,015	1,201	1,178	1,014
	Oxygen demand AT $(f = e - d)$	$kg\ O_2/d$	791	483	839	804	656
α-factor	Calculated $(g = f / e)$	-	0.50	0.28	0.41	0.60	0.55
	Measured (average)	-	0.67	0.55	0.58	0.75	0.71
	Difference	%	-25	-48	-28	-20	-22

The measured efficiency of the pre-treatment as presented in §3.2.2. has a big influence on the actual loading of the biological step. Determination of these removal efficiencies turned out not to be simple, contributing to certain inaccuracies. The reduction of the COD in the pre-treatment amounted for instance to 28% (± 15%). When it is assumed that a total COD removal during pre-treatment was 20% instead of 28%, the calculated α-factor for the total research period increases from 0.50 to 0.67 and is then equal to the measured value.

As given in the sub-study "OC and hydraulics" [3] the accuracy of the α-factor measurement is a point of discussion. Possibly the actual α-factor is lower than the measured α-factor.

Based on the oxygen balance the influence of the membrane aeration on the oxygen input in the MBR can be quantified. The results of this for the whole research period are summarized in Figure 5.7. Based on this consideration it can be concluded that the membrane tank is responsible for 33% of the oxygen input. In the membrane tank 23% of the total oxygen load is consumed. The largest part (76%) is utilised in the aeration circuit and 1% disappears with the permeate from the system.

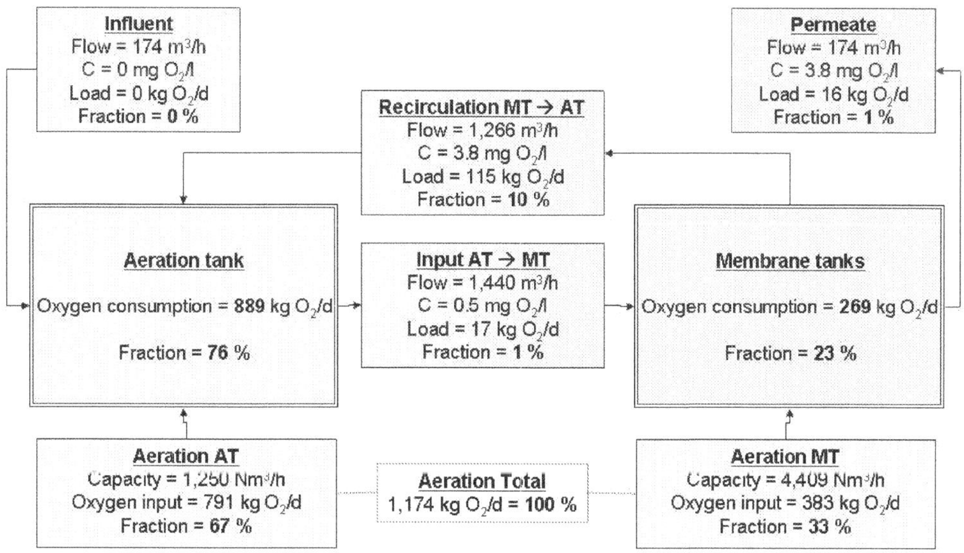

Figure 5.7. The oxygen balance for the whole research period

5.7 ENERGY CONSUMPTION

An important goal of the first year of operation of the MBR Varsseveld was to lower the energy consumption. The attention has been primarily paid to the energy use of the membrane installation, as the highest energy user, at the same time offering opportunities for optimisation. The results of it are described in the sub-study "Membranes" [3]

The decrease of the energy use cannot influence negatively on the performance of the membranes. Taking this into account the operation of the membranes was modified step by step from relatively safe (much aeration, short permeate cycle) to progressive (little aeration in stand-by mode, longer cycle, higher optimal flux when possible). The expectation for the future is to perform further optimisation steps. Hereby one can think about switching off the recycle pumps during stand-by mode and lowering of the aeration in process mode. The optimisations of the membrane installation led to an energy saving of circa 0.2 kWh/m^3. The expectations are that the energy consumption can be lowered still with 0.10-0.15 kWh/m^3. To achieve this some modification in the control installation were introduced in the middle of 2006.

The specific energy consumption of the whole installation is shown per month in Figure 5.8. Hereby the consumption is split into different main process units of the installation. Subsequently in the last four columns the specific consumption for three different periods is given and for the future situation, in which further optimisations were introduced.

The energy consumption of the whole installation was for 60-65% determined by the membrane related process sub-units. Over the whole year 2005 the energy consumption was 1.0 kWh/m^3 (or 59 kWh/p.e./year assuming p.e. of 136 g TOD) over the last seven months of 2005 it was however 0.90 kWh/m^3 (or 53 kWh/p.e./year). This decrease is the consequence of the earlier optimizations. The expectation is that at further optimisations the energy consumption could decrease to 0.75 kWh/m^3 (or 44 kWh/p.e./year).

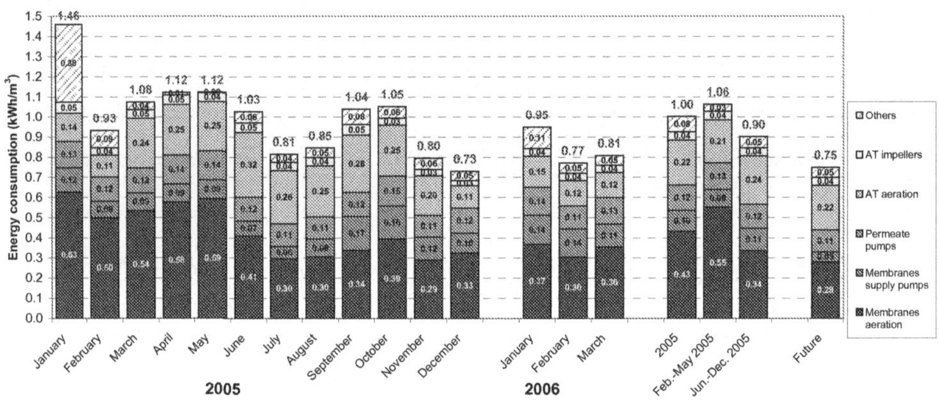

Figure 5.8 The specific energy consumption of the MBR Varsseveld

5.8 MEMBRANE OPERATION

5.8.1 Introduction

The course of the permeability in relation to the amount of the extracted permeate is shown for the four membrane tanks in Figure 5.9. Remarkable regarding the course of these graphs is the changing level of the permeability in the time. The factors which could have influence on it are described in this section.

The remarkable decrease in the permeability in the first months after the start-up is a consequence of the discharge of an industrial polymer, as described in §5.8.2. This decrease was not only related to the polymer but also to the low temperature prevailing at that moment. The effect of the temperature on the permeability is analysed in §5.8.3. During the design of the membrane tanks a lot of attention was paid to the "symmetry" of the system. The results of the related research activities are summarized in §5.8.4. Finally in §5.8.5 the presence of the EOX-compounds in the permeate as a result of the chemical cleaning is discussed.

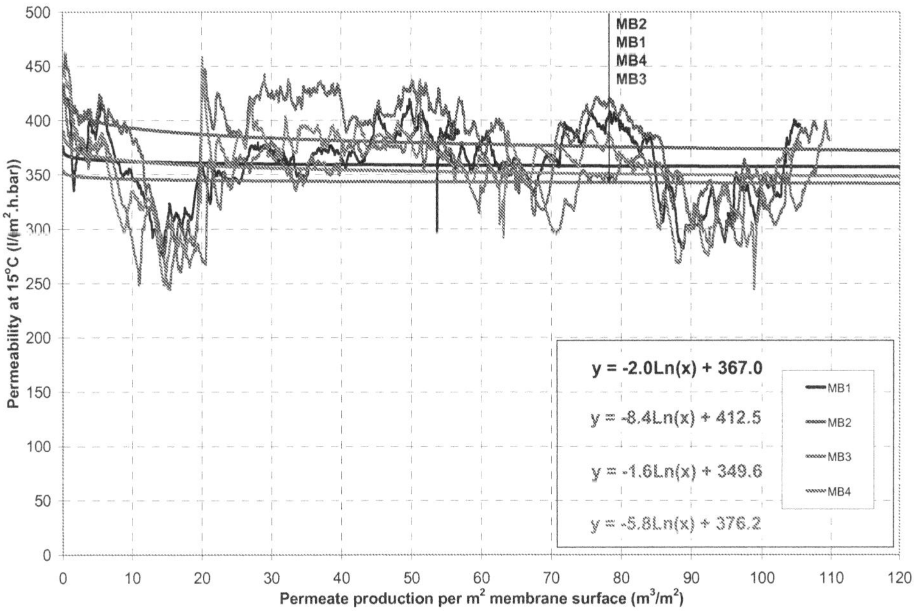

Figure 5.9. The course of the specific permeability of the four membrane tanks

5.8.2 Sludge quality

The fact that the sludge quality has a large influence on the operation of the membranes was described in detail in the sub-study "Sludge quality" [3] and §5.2.4 of this report. In the sub-study the influence of the industrial polymer on the filterability was already clearly proved. The magnitude of its influence on the working of the membranes is illustrated in this paragraph.

Under normal conditions with a good sludge quality and clean membrane, the trans-membrane pressure within a permeate cycle will not increase during the optimal flux. The optimal flux is chosen by the operator in the way that it complies with this boundary condition (see §4.4.2).

The industrial polymer caused sticking of the membranes (see Figure 4.5). The consequence of this was that a part of the membrane surface was not available for the permeate extraction and that the available surface needs to handle a much higher flux than under normal conditions. The effect of this is that the pressure increases during a permeate extraction cycle as shown in Figure 5.10. In that figure the gross flux, the trans-membrane pressure and the calculated not corrected permeability are given for a period of more than 2 hours on the 4th April 2006.

At optimal flux the membranes function as they were meant, the trans-membrane pressure is relatively constant and on the acceptable level. At increased flux the effect of the sticking became however calamitous. The trans-membrane pressure increases within a cycle of 6 minutes from circa 70 mbar to almost 500 mbar. The average gross flux amounted hereby circa 45 l/(m^2.h) but the actual flux with which the available membrane surface was loaded, was probably factors higher. The maximal allowed trans-membrane pressure for this system amounts 600 mbar. Above this measure the membrane street in question is switched off and correcting action needs to take place to recover the permeability of the membrane.

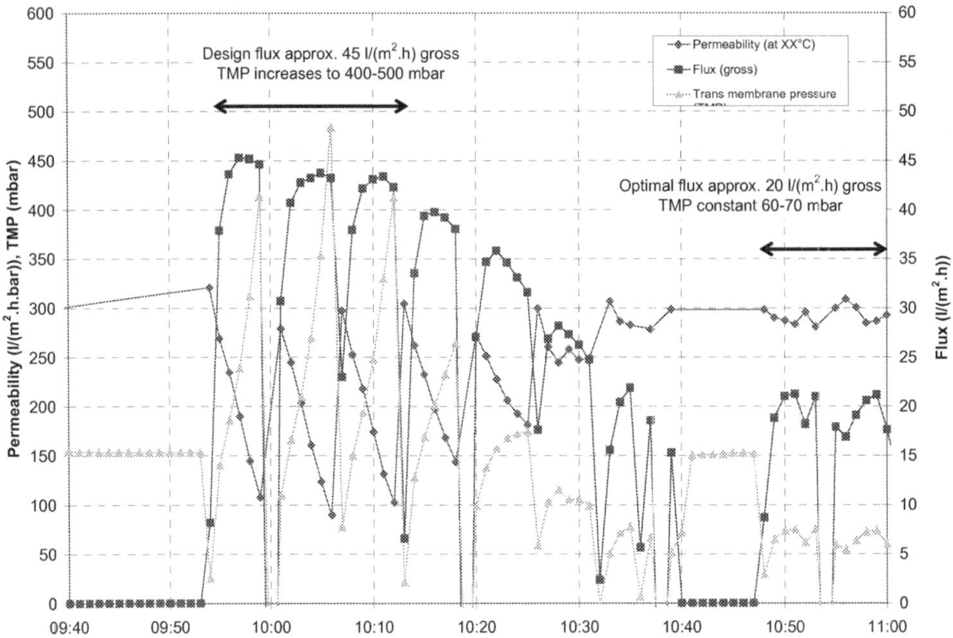

Figure 5.10. The course of the pressure increase and the permeability at presence of industrial polymer

5.8.3 Effect of temperature

The permeability depends, amongst others, on the temperature of the medium to be filtered. The temperature influences the viscosity of the activated sludge and consequently its resistance. This is therefore the rule to correct the measured actual permeability for this viscosity effect. This was also the case for the presented results of the MBR Varsseveld as shown in Figure 3.7, 3.8 and 5.9. From the results of the MBR Varsseveld however it can be observed that also in the winter the corrected permeability decreased. In 2005 this effect was attributed to the industrial polymer carrying for the sticking of the membranes. From January 2006 there was however again a clear decrease observed, from more than 350 to under the 300 l/(m^2.h.bar). Especially after the wet, cold weeks in February and March 2006 the permeability decreased more than expected. When from the middle of April the temperature starts to increase above the 15°C the permeability returns again to the old level.

In Figure 5.11 all data points are given from 15th April 2005 for one membrane tank (after the industrial polymer was removed from the system). Hereby the data points (each ½ hour) are worked out whereby the net-flux was between 15 and 20 $l/(m^2.h)$. This means that the data points during the storm weather flow were not taken into account. In the graph the corrected (for 15°C) as well as the non-corrected permeability values are shown.

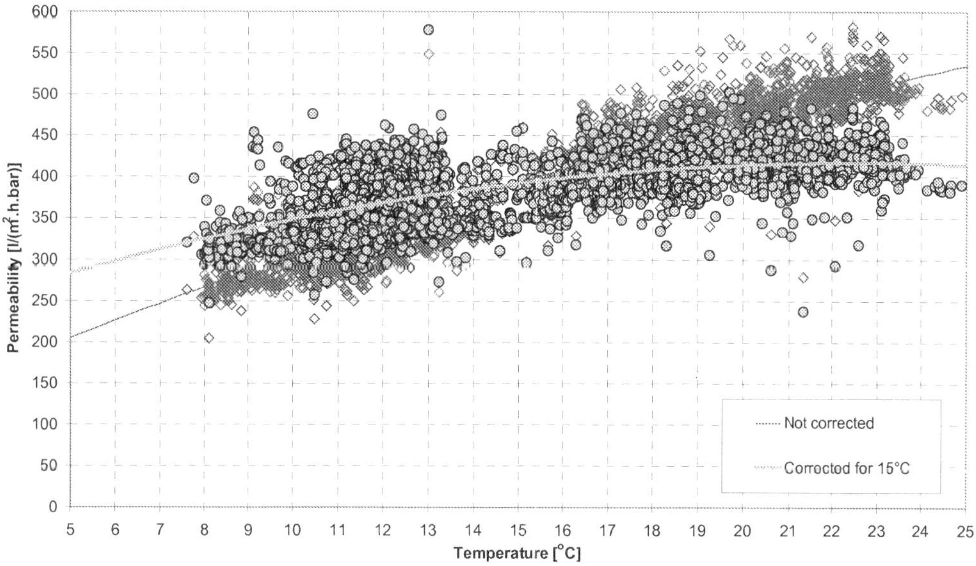

Figure 5.11. The relation between the corrected and not-corrected permeability and the temperature. (15th April 2005 – 9th May 2006, data points at net-flux of 15 to 20 $l/(m^2.h)$)

The cloud of the measurement points can be divided into three areas.

- Temperature > 15°C: the corrected permeability was almost not sensitive to the temperature;
- Temperature > 10°C and < 15°C: the corrected permeability decreases slightly at lower temperatures;
- Temperature < 10°C: the corrected permeability is lower (< 350 $l/(m^2.h.bar)$) than at higher temperatures.

It seems that below 15°C and especially below 10°C the temperature plays a bigger role than can be expected based on the viscosity effect. The reason for this phenomenon was not further investigated, but it can be related to the following aspects:

- The viscosity of activated sludge (at 10 kg $MLSS/m^3$) changes more strongly than the viscosity of water; therefore the applied viscosity correction is not then sufficient;
- The activated sludge contains relatively high amounts of greasy material which becomes more viscous at lower temperatures;
- The membrane properties change causing the increase of the membrane;

- The degassing of the permeate in the permeate pipe increases causing an increase of the resistance.

5.8.4 Symmetry

The MBR Varsseveld is the first large scale MBR installation in the Netherlands. The experience with the MBR technology was based up to now on the research with pilot installations with a capacity of 2 to 20 m^3/h. The capacity of the MBR Varsseveld is approximately 40 times larger than that of the largest pilot installation. The membrane tank such as this one built in Varsseveld serves as a model for a standard membrane tank in large scale MBR implementations. A starting point during the design was a complete symmetry of the system. This means that the conditions for all membrane cassettes in a tank are the same. This means that the influent and effluent of the activated sludge are equally distributed in each membrane tank.

In the sub-study “Membranes” [3] much attention was paid to determine the symmetry of the membrane tanks. To achieve this, the measurements were performed of the so called filtration balance and the chemicals’ distribution. The hydraulic division in a membrane tank was investigated in the sub-study “OC and hydraulics” [3]. The results of both researches are summarised here.

Filtration balance

During the determination of the filtration balance the operation of each membrane cassette was investigated and the differences between the cassettes were evaluated. During the start-up of the membrane installation the permeability of clean water was determined (Clean Water Permeability, (CWP)) of the different cassettes. After approximately 10 months of operation the filtration balance in activated sludge was determined in September 2005.

The results show the differences between the membrane tanks and the membrane cassettes. The similarities between the measurements in clean water and activated sludge are remarkable. The permeability after 10 months operation seems to be strongly related to the original CWP-value of the membrane cassettes. This suggests strongly that the differences between the CWP-values of the cassettes result from the variations in the product quality. By changing the set-ups in the production process the quality, and consequently the permeability, of the membrane changes.

Distribution of the chemicals

On the MBR Varsseveld a maintenance cleaning is implemented whereby permeate with a solution of chemicals is pumped back to the membranes. An equal distribution of the cleaning chemicals over all membrane elements is of large importance for a stable operation of the membranes. When this is not the case, one element will be cleaned to a worse extent than another and an imbalance occurs in the membrane installation. To investigate this phenomenon, the measurements were performed into the chemicals distribution between the cassettes within one street and between the elements within one cassette.

The hydraulic distribution between one cassette shows that in the middle of the cassette relatively more chemicals are washed back than on the side. This distribution is more or less similar for all cassettes; there are not significant differences to be seen between the first and the last cassette within one tank.

In Figure 5.12 the results of the filtration balance tests and the chemicals distribution tests (distribution between the cassettes within one tank) are plotted in one graph. From this relation it can be seen that the hydraulic distribution during the back wash is strongly related to the filtration balance. The distribution of the chemicals during the backwash seems to be mainly influenced by the permeability of the considered cassettes.

Hydraulic distribution

During the design of the membrane tank the attempt was taken to achieve as equal as possible a distribution of the sludge influent and effluent, in the way that no or insignificant concentration gradients occurred in a membrane tank. Gradients in, for instance, dry matter content, can negatively influence the membranes. Hydraulic tests were performed to measure the liquid retention time distribution in the membrane tank. Activated sludge was used here as a tracer component. Based on the results of these measurements it can be stated that the membrane tanks can be considered as a completely mixed tank.

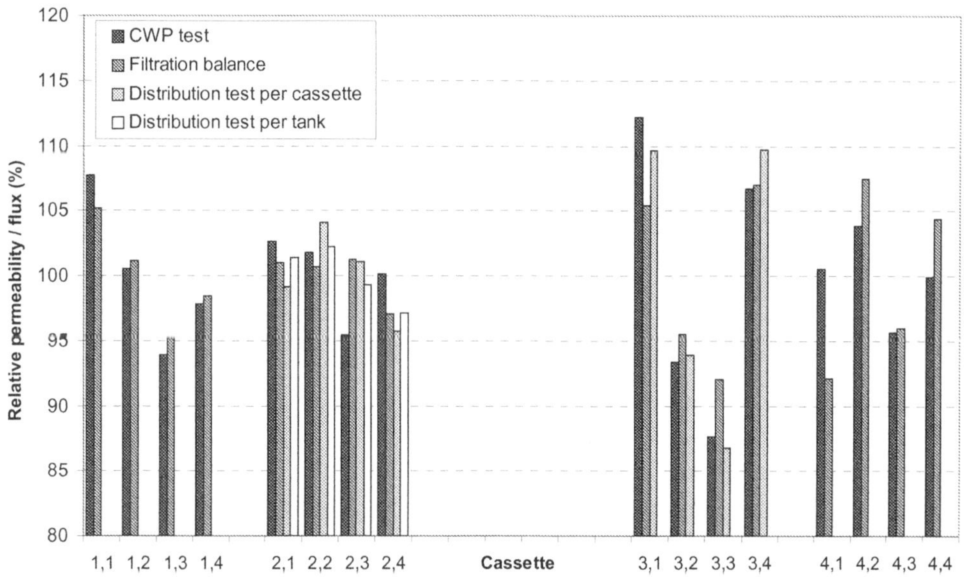

Figure 5.12. Distribution between the cassettes based on the different measurement methods

EOX forming

The membranes are periodically cleaned with sodium hypochloride and citric acid. From the environmental unfriendly sodium hypochloride chlorinated organic compounds are formed. Within the sub-study "Specific micropollutants" [3] this was investigated by measuring the

EOX permeate content. The measurements were performed just before the cleaning and 10, 20 and 60 minutes after the cleaning. In Table 5.8 the results are presented.

Tabel 5.8 Results of the EOX measurements (in µg/l) in the permeate just before and after the cleaning of the membranes

Sample taking time	0 minutes	10 minutes	20 minutes	60 minutes
4th April 2006	<1	<1	<1	1
11th April 2006	<1	<1	<1	2

The EOX concentrations were low, maximally 2 µg/l, corresponding to the earlier results attained by the Water Board. The results do not correspond however with the earlier measurements on the MBR pilot-installation in Maasbommel, the Netherlands [9]. There, directly after the chemical cleaning an average EOX-concentration in the permeate of 11 µg/l was measured. These differences are possibly a result of the applied cleaning procedure and a too short sampling period.

On the MBR-pilot plant of Maasbommel the cleaning was performed in medium. After the cleaning permeate was directly extracted from the medium that was polluted with rest chlorine, causing the relatively high and directly measurable EOX-peak. On the MBR Varsseveld a chemical cleaning procedure is performed in air, meaning in a membrane tank filled only with a little amount of activated sludge (circa 10 m^3). After the first cleaning step with sodium hypochloride the largest part of the present activated sludge, including the chemicals is present in there. Subsequently once more the same amount of activated sludge is brought into the membrane tank and the second cleaning step is performed with citric acid. After this the sludge was also discharged, the membrane tank was filled with activated sludge from the aeration tank and put in operation.

The present rest chlorine is in this way almost completely recycled to the aeration tank where it is possibly converted to chlorinated compounds. The fact that these were not detected can be explained by a too short sampling period. Before the formed EOX reaches permeate it has to be brought via the pre-denitrification tank and the aeration circuit to the membrane tanks. The average hydraulic retention time in the pre-denitrification tank and aeration circuit amounted respectively to 20 and 60 minutes. A possible EOX-peak is therefore to be expected after 80 minutes.

It is remarkable to see from the measurements shown in Table 5.8 that after 60 minutes the highest EOX-concentration in the permeate was measured. Hereby it has to be reported that these values were not higher than the measurements on days when no chemical cleaning took place.

For the time being it can be concluded that cleaning with hypochlorine did not lead to increased concentrations of chlorinated compounds in the permeate of the MBR installation. To have a definitive answer about this, complementary measurements are required for a longer period of time.

6

Costs comparison

6.1 INTRODUCTION

In this section the costs of the MBR Varsseveld and a conventional WWTP with sand filtration as a polishing step are compared with each other. The investment costs are presented in §6.2. The actual costs of the MBR Varsseveld were used here. The exploitation costs of both technological options are calculated §6.3. An overview of the investment- and operational costs are given in §6.4.

6.2 INVESTMENT COSTS

The calculation of the investments costs of the MBR Varsseveld are based on the real costs. For the determination of the investment costs of the reference alternative the feasibility study results for the MBR Varsseveld were used and the cost comparison set-up for the research on the MBR Maasbommel [9].

In 2000 in the feasibility study there was a comparison made between the conventional extension of the existing WWTP Varsseveld, including the implementation of sand filtration, and the building of an MBR. Based on these cost calculations an MBR was chosen then, despite that this option was somewhat more expensive (circa 5%). During the assessment of the conventional enlargement, the partial reuse of the existing units was assumed. When it was not the case the costs of the MBR and conventional extension would be almost the same.

In the STOWA report over the MBR Maasbommel the comparative investment costs for an MBR and a conventional activated sludge system with sand filtration were determined. Hereby a zero situation was assumed for both variants at two different loading situations (10,000 and 50,000 p.e.). At the 10,000 p.e. the MBR installation turned out to be circa 9% cheaper than the reference variant. At the 50,000 p.e. comparison, the MBR was 1% more expensive.

Based on both studies it can be concluded that for an installation of 10,000-50,000 p.e. the costs for the MBR and the reference variant are almost the same. This statement was used as a starting point in the cost comparisons. The division of the construction costs in the different sub-parts (civil, mechanical engineering, electric, membranes) showed large variations. This was based on the division as determined in both mentioned studies. The result of the cost comparison between MBR and the reference variant is given in Table 6.1. The starting costs were based on the actual start costs for the building of the MBR Varsseveld.

Table 6.1 The investment costs of the MBR Varsseveld and the conventional enlargement

Expenses		Unit	MBR Varsseveld	Reference variant
Building costs	Civil	€	2,850,000	5,250,00
	Mechanical	€	1,710,000	1,570,000
	Electric/automation	€	1,280,000	1,440,000
	Membrane system[1)]	€	2,430,000	-
	Total	€	8,260,000	8,260,000
Initial costs[2)]		€	2,940,000	2,940,000
Investment costs		€	11,200,000	11,200,000

1) Incl. permeate pumps, blowers, pipes, etc.

2) Including staff costs, consulting costs, building inspection and tax

6.3 OPERATIONAL COSTS

6.3.1 Introduction

The estimate of the operation costs was based on the actual costs of the MBR Varsseveld. In the cost estimate, the consumption of the resources (energy, chemicals), the costs of the sludge disposal and the costs of the maintenance of the installation (man power) were included. Next to it the capital expenses were taken in the comparison, assuming annual depreciation.

The starting points used in the costs estimation are given in Table 6.2.

Table 6.2 Starting points operational costs

Description		Unit	Value
Interest		%	4.5
Depreciation period	Civil	Year	30
	Mechanical /electric	Year	15
	Membranes	Year	8
Maintenance	Civil	%	0.5
	Mechanical/electric	%	2.0
Energy		€ / kWh	0.075 [1,2)]
Iron chloride sulphate		€ / ton product	114 [1)]
Sludge disposal (incl. transport)		€ / ton MLSS	550 [1)]
Staff		€ /person	40,000

1) Excluding tax

2) Including regulating energy tax, standing charge and variable costs

6.3.2 Energy consumption

The average energy consumption of the MBR Varsseveld amounted in 2005 1.0 kWh/m^3. As a result of different optimisations it was in the last months of 2005 and the first months of 2006 reduced to 0.88 kWh/m^3. The expectation is that based on further optimizations the energy consumption will decrease to 0.75 kWh/m^3 (see §5.7). In the comparison it was however assumed that the actual energy consumption was 0.88 kWh/m^3. The energy use of the reference variant was taken from similar systems (WWTPs Ruurlo and Wehl of the Water Board Rijn en IJssel, research Maasbommel [9]) and amounts to 0.65 kWh/m^3.

The cost comparison for the energy use is summarised in Table 6.3. As a starting point the average influent flow over the period of one year from March 2005 until March 2006, which was 4,320 m^3/day.

Table 6.3 Energy costs

Option	Energy use		Specific costs	Costs
	KWh/m^3	MWh /year	€ / kWh	€ /year
MBR	0.88	1,388	0.075	104,000
Conventional with sand filters	0.65	1,025	0.075	77,000

6.3.3 Chemicals consumption

To remove phosphate iron chloride sulphate was dosed. When a phosphate permeate content of 0.15 mg P_{total}/l has to be achieved an ferric dosing of circa 1.2 mol Fe/mol $P_{influent}$ is needed (see Figure 3.4). This corresponds with a dosing of circa 540 l/day. From November till March a permeate concentration was however 1.0 mg P_{total}/l. This corresponds with circa 0.6 mol Fe/mol $P_{influent}$, or 270 l/day. The iron salt use over the whole year was therefore 156 m^3. The starting point was that for the reference variant the same amount of iron salt was needed.

The costs of the iron chloride dosing for both options are given in Table 6.4.

Table 6.4 Costs of ferric dosing

Option	$FeClSO_4$-dosing		Specific costs	Costs
	m^3/year	*ton product /year*[1)]	*€ / ton product*	*€ / jaar*
MBR	156	231	114	26,000
Conventional with sand filters	156	231	114	26,000

1) Density is 1,500 kg/m^3, 41% solution

For the membrane cleaning of the MBR the use of chemicals is necessary. On the MBR Varsseveld sodium hypo chloride and citric acid were used to achieve this goal. The yearly costs of this amounted to € 10,000, as reported in §4.2.4. of the sub-study "Membranes" [3]. For the reference variant no additional chemicals are necessary.

6.3.4 Sludge management

The sludge production of the MBR Varsseveld and of a comparable conventional system are calculated in Table 5.3. For the sludge transport of the MBR the transport of sieve goods is included.

The costs of sludge disposal are given in Table 6.5.

Table 6.5 Costs of the sludge disposal

Option	Sludge disposal		Specific costs	Costs
	kg MLSS/day	*tone MLSS/year*	*€ / tone MLSS*	*€ /year*
MBR	1,360	496	550	273,000
Conventional with sand filters	1,400	511	550	281,000

6.3.5 Staff costs

The Water Board Rijn en IJssel made an estimation of the staff costs on the MBR Varsseveld in the first quarter of 2006. This was compared to the staff expenses on a conventional WWTP with sand filtration (Ruurlo) of a similar size. The additional staff costs on the MBR Varsseveld are estimated to be 50%.

The results of this comparison are given in Table 6.6.

Table 6.6 Staff costs

Option	Staff	Specific costs	Costs
	persons	*€ / person*	*€ / year*
MBR	0.9	40,000	36,000
Conventional with sand filters	0.6	40,000	24,000

6.4 OVERVIEW

The investment and the exploitation costs of the MBR Varsseveld and the reference variant are summarised in Table 6.7. The exploitation costs of the MBR Varsseveld amount to 1,640,000 €/year. Of this 60% constitutes the depreciation of the installation and the membranes. Of the remaining, 40% are operational costs and 40% is spent on the sludge disposal. The costs for the chemical membrane cleaning amount to 1.5% of the total operational costs.

Table 6.7 Overview of the investment- and exploitation costs

Costs		Unit	MBR Varsseveld	Reference variant	Difference
Investment		€	11,200,000	11,200,000	-
Exploitation costs	Total (1 + 2, round off)	€ / year	1,640,000	1,400,000	+ 240,000
Operation	Total (1)	€ / year	658,000	590,000	+ 68,000
	Energy	€ / year	104,000	77,000	+ 27,000
	Iron chloride sulphate	€ / year	26,000	26,000	-
	Chemicals membranes	€ / year	10,000	-	+ 10,000
	Sludge disposal	€ / year	273,000	281,000	- 8,000
	Staff	€ / year	36,000	24,000	+ 12,000
	Maintenance	€ / year	110,000	92,000	+ 18,000
	BTW	€ / year	99,000	90,000	+ 9,000
Capital expenses	Total (2)	€ / year	984,000	811,000	+ 173,000
	Depreciation installations	€ / year	801,000	811,000	- 10,000
	Depreciation membranes	€ /year	183,000	-	+ 183,000

In Table 6.8 the specific exploitation costs are presented, expressed per p.e. (a 136 g TOD) and per m^3 treated wastewater.

Table 6.8 Specific operational- and exploitation costs

Cost specification	Unit	MBR Varsseveld	Reference variant
Exploitation costs	€/m^3	1.07	0.92
	€/(p.e./year)	64	54

7

Evaluation

7.1 INTRODUCTION

In this report the design, the results and the experiences of the first 16 months of operation of the MBR Varsseveld were presented, as well as the results of the research program. In this last chapter the design and the performance of the MBR Varsseveld are evaluated based on the goals as formulated in §1.2. These are discussed in the following sections:

- Determination of the technological results and optimisation of the process operation (§7.2);
- Proving a stable operation (§7.3);
- Evaluation of the design and research into the necessity of redundancy (§7.4);
- Comparison of the costs of the MBR against a conventional WWTP (§7.5);
- Demonstration of the technical feasibility of the scaling-up (§7.6).

7.2 RESULTS

The biological results of the MBR Varsseveld are good. The relatively stringent target value for nitrogen removal (5 mg N_{total}/l) was easily reached. This was caused, by amongst others, the relative low nitrogen loading and the relatively high COD/N-ratio of the wastewater. The target value for phosphate removal (0.15 mg P_{total}/l) was not reached for the larger period of time because this was not the goal of the process operation. Phosphate removal was optimized by stepwise increasing the iron salt dosing. Finally it led to a period of almost 1½ months when the phosphate requirement was reached at a dosing of circa 1.2 mol Fe/mol $P_{influent}$. Despite the absence of an anaerobic tank biological phosphate removal played an important role. The necessary chemical consumption was therefore, despite the relatively high influent phosphate concentration, relatively low.

The permeate results showed that the reaching of the MTR- standard for phosphate (0.15 mg P_{total}/l) on the WWTP Varsseveld was not possible without chemical dosing. The iron salt dosing resulted in a decrease of the soluble organically bound phosphate content from circa 0.5 to 0.03 mg P/l. The influence of the organically bound phosphate content on the feasibility of reaching the effluent requirement is different per location, and depends mainly on the influent composition and the sludge loading. A relatively small dosing of iron salts can lead to a significant decrease of the organic phosphate concentration and therefore it can enable the permeate quality requirement to be reached more easily.

The membrane installation has always functioned well, with the exception of the first few months when the discharge of an industrial polymer disturbed the operation of the installation. After the operation efficiency of the membranes was recovered, by the intensive cleaning of the membranes and stopping the discharge of the polymer, the installation could handle the influent load under all conditions. Under relatively beneficial conditions (high process temperature) the guarantee tests for the membranes were successfully performed.

The energy use of the membrane installation amounted circa 60% of the total energy use. There was therefore much attention paid to the lowering of the energy consumption. Due to this the energy consumption decreased from 1.0 kWh/m^3 (or 59 kWh/p.e./year, assuming p.e's of 136 g TOD) in 2005 to 0.88 kWh/m^3 (or 52 kWh/p.e./year) in the period of June 2005 to March 2006. The average energy use of a comparable conventional WWTP with effluent filtration amounted approximately to 0.65 kWh/m^3 (or 38 kWh/p.e./year) (see §6.3.2). The energy consumption of the MBR Varsseveld was therefore 35% higher than that of a conventional activated sludge sand filtration. The expectation is that, after some adjustments of the control installation, the energy use can be decreased further to circa 0.75 kWh/m^3 (or 44 kWh/p.e./year). After this optimization the energy use of the MBR Varsseveld would be circa 15% higher than that of a conventional activated sludge with sand filtration.

7.3 OPERATION

The operation of an MBR differs strongly in many points from a conventional WWTP. This comprises primarily the membrane filtration installation. This results not only in a completely new type of process operation but also a number of new process sub-units. The expectation that especially these new sub-units would lead to problems, was not justified.

The membranes did function excellently for the largest part of the time, regarding both, technological as well as technical aspects. The technical problems which occurred with leakages were the result of the scaling-up of the MBR-technology and were solved efficiently by the manufacturer. The macro-fouling which occurred was most probably the result of the too low aeration capacity during the stand-by mode and the sub-optimal way of aeration (sequentially instead of air-cycling), in combination with a problem with the pre-treatment. The expectation is that this problem can be solved after some technical adjustments and further optimisations.

The experiences with the industrial polymer are an important lesson regarding the design of future MBR installations. Influent components, which did not ever lead to any problems in conventional installations, can do so in an MBR installation. Components, which cannot pass the membranes, accumulate in the system and can lead to unexpected problems. In

Varsseveld it was an industrial polymer, used in the local cheese factory as a cheese coating agent, which caused large problems with the membranes. The same agent is however applied in several other food industries and sectors. A critical analysis of the wastewater and of industrial discharges in the considered area is recommended before the decision is made for the implementation of an MBR. Hereby the main attention should be paid to polymers, oils and grease, inert eroding particles (such as fine sand), solvents (also harming biology) and scaling components (calcium, magnesium). The experience in Varsseveld showed that even the performance of the pilot-research does not automatically lead to spotting these kinds of problematic compounds.

An efficient pre-treatment is essential for a reliable membrane performance. The pre-treatment, and especially the micro-sieves, was the sub-unit at the MBR Varsseveld where the most attention was paid. Despite the standard application of the complete 100% redundancy, the problems with clogging of the micro-sieves are not (yet) completely solved.

The failing of the pre-treatment can disable a successful development of the MBR-technology. There is therefore more development needed in the area of pre-treatment. Hereby the experiences of users and manufactures should be combined.

7.4 DESIGN

The pre-treatment seems to be the most disturbing part of the installation. Especially the micro-sieves which cannot always handle the influent. This sensitivity suggests that the 100% redundantion of the micro-sieves, as implemented at the MBR Varsseveld, is recommended. The fine screens operated always without any nuisance. The redundancy of this process step based on the first 16 months of operation did not turn out to be essential. Regarding the importance of the fine screens for a stable operation of micro-sieves, their 50% redundancy is recommended. The choice of a finer mesh can be considered (for instance 3 mm instead of 6 mm), to reduce the loading of the microsieves.

Based on the results of the complementary measurements it can be assumed that no grease is removed in the grit/grease removing chamber. The grease is filtered from the wastewater for a large part in the fine screens and the micro-sieves. Although in practice the presence of grease was observed in the grease drain, the presence of the grease chamber does not seem to be essential for the good operation of the MBR. The necessity of a separate grease chamber needs to be further investigated. A separate sand removal is recommended to minimize damage to the process units' parts and the membranes. The redundancy is not necessary here.

The biology of the MBR Varsseveld consists of an aeration circuit with bubble aeration and a pre-denitrification tank. The choice of such a system was determined by the relatively stringent effluent nitrogen requirements of 5 mg N_{total}/l. To attain such low nitrogen concentrations a sufficient sludge recycle and frequent alternation between oxic and anoxic zones are of importance. Thanks to, amongst others, an optimal system configuration, the results of nitrogen removal are very good.

Based on the measurements of the oxygen input and the hydraulics of the aeration circuit, it can be concluded that the aeration circuit is not optimal for this kind of application. Because a relatively large volume of air needs to be brought in a small volume, the available volume for the aeration elements and impellers is very restricted. This led to a sub-optimal design whereby the necessary space between the aeration and impellers on one site and aeration/impellers and the curves in the tank on the other site, was smaller than normal. As a consequence of this, the aeration circuit hydraulically does not function optimally; the flow velocities are too low and the oxygen input capacity is lower than desired. In addition, the design guidelines for the applied plate aeration turned out to be too optimistic. These factors led to an oxygen input capacity 30% lower than it should be according to the design.

Based on the oxygen balance it can be concluded that the membrane aeration at the MBR Varsseveld takes 20-25% of the oxygen input. Although in the membrane tanks more than 3 times more air is provided than in the aeration circuit, the oxygen input is 3 times lower.

In the safe design of the MBR Varsseveld, the efficiency of the pre-treatment step for all components was set to 0%. It is recommended to assume the following design starting points when designing an MBR with fine screens and micro-sieves (depending on the type of micro-sieve and the mesh):

- COD/BOD/ Suspended solids removal : 20 – 30%;
- N_{Kj} / P_{total}-removal : 5 – 10%.

The MBR seems to be more sensitive to a scum layer forming than conventional activated sludge. In the design this has to be taken into account. The following measures are recommended:

- A scum layer drain in the aeration tank;
- Additional warning level in the aeration tank;
- A scum layer sprinkling installation;
- Sub-surface location of the on-line analysers (not floating);
- The influent to the membrane tanks needs to take place from a few metres below the surface.

Based on the guarantee tests it can be concluded that the membrane filtration installation did fulfil the design requirements. This comprises a design flux of 37.5 l/(m^2.h) during 3 days followed by a maximum flux of 50 l/(m^2.h) during 8 hours. The guarantee tests were however performed in the summer at the beneficial process temperature. In the winter the permeability seemed to decrease faster during the peak loadings than expected. Based on the available data it was difficult to predict whether the guarantee test would also be completed successfully in the winter. At this moment it is not recommended, for the design of the following MBR with the same types of membranes (Zenon ZW500d) to have a design flux higher than that of the MBR Varsseveld. The design flux is strongly related to the duration of the maximal peak factor.

The aeration of the membrane tanks is essential for a good and stable operation of the installation. Since the application of air-cycling caused technical problems in Varsseveld it was switched to the sequential aeration. The efficiency of air-cycling is expected to be higher; with the same amount of air possibly cleaner membranes can be maintained. The expectation is that implementation of air-cycling can minimise the occurrence of macro-fouling, especially the sludge accumulation on the lower side of the membranes. The implementation of air-cycling is to be investigated further on the MBR Varsseveld in 2006.

The scaling up of the membrane tanks to units, which can handle maximally 250 m^3/h was successfully accomplished. The symmetry of the system was proved and the performance of the membranes was excellent. The choice of the number of membrane tanks in Varsseveld was based on the operational and technical requirements. From the operational point of view it is required to always handle the maximal loading, also when the technical disturbances disable operation of one of the membrane tanks. In the installation with 4 membrane tanks 3 remaining tanks need to operate and handle 4/3rd of the design flux. These temporary increases in the flux of 33% are possible for membranes, which are well maintained. Next to it, during the design (2001-2002) the maximally available capacity of the permeate pumps with variable flow direction was circa 300 m^3/h. Meanwhile larger capacities are available on the market and larger membrane tanks with similar permeate pumps can be realised. A further optimisation for future design comprises the building of more compact membrane tanks, increasing the efficiency of the aeration.

7.5 COSTS

The investment costs of the MBR Varsseveld are approximately similar to the conventional extension with sand filtration. The exploitation costs (operational costs and capital expenses) of the MBR Varsseveld are approximately 18% higher than the reference variant.

Based on these cost comparisons it can be concluded that the cost difference between the MBR and the conventional WWTP with sand filtration is relatively small. Future developments (lower energy use, lower membrane prices) will possibly bring the exploitation costs of both variants to the same level.

7.6 TECHNICAL FEASIBILITY

Based on the results and experiences of the first 16 months of operation on the MBR Varsseveld it can be concluded that the MBR technology is applicable for the treatment of the communal wastewater in the Netherlands. The results of the installation were in general very good. The target values of the permeate quality were reached and the membranes functioned excellently (except for some technical problems). The investment costs of the MBR are in the meantime similar to those of a conventional installation with sand filtration and also with respect to the energy consumption the differences are not too large. To make MBR-technology suitable for future large scale applications it is important to solve the critical bottle necks and to supplement with the most important knowledge gaps.

The two most important critical factors determining the success of the MBR are the pre-treatment and the influent composition. The problems with the micro-sieves on the MBR Varsseveld need to be solved to provide a stable operation. The further development of a new generation of micro-sieves should lead to robust and reliable equipment. In addition more information should be gained with respect to the sensitivity of the membranes to the components present in the wastewater. In the design stage of a new installation it has to determined in advance whether the components present in wastewater can disturb the operation of the membranes. The experiences on the future installations in Heenvliet, Ootmarsum and Hilversum should provide more insight into both these critical aspects.

Finally it is of importance to perform more fundamental research. The increase in knowledge will lead to more insight into the processes, better design and therefore better functioning MBR-installations. The most important aspects to be further researched are:

- The effect of the process temperature on permeability (namely at temperatures lower than 10°C);
- The effect of the sludge quality on the filterability, as a continuation of the sub-study “Sludge quality” [3]. Finally it should lead to design guidelines for the optimal design guidelines for membranes and for the optimal biology.

REFERENCES

[1] STOWA (2002A) MBR for municipal wastewater treatment – pilot plant research Beverwijk WWTP. STOWA report 2002-11A.
[2] STOWA (2002B) MBR for municipal wastewater treatment – Pilot plant research Beverwijk WWTP. Side-studies, STOWA report 2002-11B.
[3] STOWA (2006) Research MBR Varsseveld, sub-study report, STOWA sub-study report 2006-06 (in Dutch).
[4] STOWA (2005) Monitoring of endocrine disrupting compounds and pathogens on WWTP with supplementary treatment techniques. STOWA report 2005-32 (in Dutch).
[5] Seyfried, C.F., Schüssler, H., Lohse, M., en G. Bebendorf. (1985) Comparison of treatment efficiencies of screens, sieves and micro-sieves and their influence on further removal efficiencies. Publications of the Institute of Water Quality and Waste Management of University of Hannover, Vol. 58 (in German).
[6] STOWA (2001) Manual of biological phosphate removal. STOWA 2001-15.
[7] STOWA (2006) MBR pilot-plant Hilversum, STOWA report 2005-16 (in Dutch).
[8] Cornel P., Wagner M., Krausse S., (2002) Investigation of oxygen transfer rates in full scale membrane bioreactors. Darmstadt University of Technology, Institute WAR, No. e21291a
[9] STOWA (2004) Comparative research between MBR and sand filtration WWTP Maasbommel. STOWA report 2004—28.

APPENDIX 1 – GLOSSARY

Term	Symbol	Unit	Description
Air-cycling			Alternating aeration between the cassettes
Alfa-factor / α-factor	α	[-]	The ratio between the oxygen diffusion in the activated sludge and clean water
Back-pulse			Periodical flow change of the permeate stream through the membrane with a goal to remove the fouling of the membrane surface and of pores.
Flux	F	[l/(m^2.h)]	The amount of permeate led through a membrane surface in the time interval
Flux - Gross	F_{gross}	[l/(m^2.h)]	The actual flux during permeate extraction
Flux - Maximum		[l/(m^2.h)]	The net flux at the maximal hydraulic loading, when 1 membrane tank out of operation (= 50 l/(m^2.h))
Flux - Minimum		[l/(m^2.h)]	The net flux at the minimum capacity of the permeate pump (≈ 10 l/(m^2.h))
Flux - Net	F_{net}	[l/(m^2.h)]	The average permeate over a longer period, whereby the production decrease is taken into account as a consequence of the back-pulse
Flux - Design		[l/(m^2.h)]	The net flux at the maximal hydraulic loading, when all membrane tanks are in operation (= 37,5 l/(m^2.h))
Flux - Optimum		[l/(m^2.h)]	The net flux whereby the membranes function optimally (≈ 20 l/(m^2.h))
Hydrophobic			Water repulsive. Opposite to hydrophilic
Intensive Cleaning	IC		Cleaning method whereby the whole membrane tank is filled with cleaning solution with a goal to return the original permeability of the membrane

Term	Symbol	Unit	Description
Macro-fouling			Fouling of the membrane caused by coarse particles from the wastewater (such as hair, fibres), the activated sludge (for instant anaerobic sludge agglomerates) or from the surrounding (for instance falling leaves)
Maintenance Cleaning	MC		A cleaning procedure, consisting of an oxidative and an acid cleaning, whereby an chemical solution of low concentration is back washed through the membranes in a (partially) empty membrane tank. This cleaning is of a preventive nature.
Membrane			A filter with small pores, applied, amongst others, to separate activated sludge from purified effluent.
Membrane bioreactor	MBR		Suspended activated sludge system whereby the separation of activated sludge and the treated effluent takes place with membranes instead of secondary clarifiers
Membrane cassette			Unit consisting of more membrane elements. In Varsseveld each cassette consisted of 40 elements. The membrane surface of a cassette was 1,260 m^2
Membrane element			The smallest membrane unit consisting of membranes and a permeate collecting header. In Varsseveld each element has a membrane surface of 31.5 m^2
Membrane tank			Space filled with the membrane cassettes. In Varsseveld there are 4 membrane tanks each with 4 membrane cassettes. Per membrane tank a membrane surface of 5,040 m^2 is installed

Term	Symbol	Unit	Description
Maximal Tolerable Risk	MTR		A minimum water quality defined in the Fourth National Decree of Water Management. For nitrogen and nitrogen phosphate the MTR standard is set to respectively 2.2 mg N_{total}/l and 0.15 mg P_{total}/l
Permeate			Term for the product flowing through a membrane and discharged from the system: the effluent of a membrane installation.
Permeate cycle			Cycle when subsequently permeate extraction and permeate back wash (back-pulse) takes place.
Permeability (at the actual temperature)	P (XX°C)	[l/(m^2.h.bar)]	The actual gross flux divided through the trans-membrane pressure (TMP) over the membrane. $P = F_{gross} / TMP$ The permeability of a membrane is a measure of the resistance of the membrane to water, so it does flow through the membrane surface under influence of the driven force (TMP), exposed onto the water.
Permeability, Corrected	P (15°C)	[l/(m^2.h.bar)]	The standardised permeability, expressed at a temperature of 15°C. It considers a correction for the increase of the viscosity at lower temperatures.
Process mode			Process set-up whereby the permeate extraction, the aeration and the recycle pumps are in operation.
Sequential aeration			Alternating aeration within a cassette
Semi-Intensive Maintenance Cleaning	SIMC		An MC cleaning with an additional oxidative cleaning (NaOCl)

Term	Symbol	Unit	Description
Simulation unit			Pilot installation
Sludge age		[d]	The amount of sludge divided through the amount of sludge removed regularly as an excess sludge or with effluent. It considers the time needed for a complete renewal of the sludge.
Sludge loading		[g COD/(g MLSS.d)]	The amount of pollution imposed onto the unit of mass of activated sludge in time.
Stand-by mode			A mode in the process set-up where no permeate extraction takes place. The aeration (discontinuous) and the recycle pump are then in operation.
Trans-membrane pressure	TMP	[mbar or kPa]	The pressure fall over the membrane, between the activated sludge side and the permeate side. It is the driving force causing the filtration through the membrane (10 mbar = 1 kPa)

IWA Publishing's authorised EU representative for General Product Safety Regulations is Diane D'Arras, 15 rue Duret, 75116 Paris, France, e-mail: safety@iwap.co.uk.

Printed and bound by CPI Group (UK) Ltd, Croydon, CR0 4YY
16/06/2026
02140370-0001